R. H Otter

Winters Abroad

Some Information Respecting Places Visited by the Author on Account of his Health

R. H Otter

Winters Abroad

Some Information Respecting Places Visited by the Author on Account of his Health

ISBN/EAN: 9783337140014

Printed in Europe, USA, Canada, Australia, Japan

Cover: Foto ©berggeist007 / pixelio.de

More available books at **www.hansebooks.com**

WINTERS ABROAD.

SOME INFORMATION RESPECTING

PLACES VISITED BY THE AUTHOR ON ACCOUNT OF HIS HEALTH.

Intended for the Use of Invalids.

AUSTRALIA:
   MELBOURNE.
   TASMANIA.
   SYDNEY
   QUEENSLAND.

THE RIVERINA.
ALGIERS.
EGYPT.
CAPE OF GOOD HOPE.
DAVOS.

By R. H. OTTER, M.A.

LONDON:
JOHN MURRAY, ALBEMARLE STREET.
1882.

[All Rights reserved.]

LONDON:
BRADBURY AGNEW, & CO., PRINTERS, WHITEFRIARS.

PREFACE.

My object in writing this book is to give those persons who are advised by their doctors to spend their winters away from England on account of health, some information respecting the different places which I have myself visited for the same reason.

I do not purpose giving any description of these places beyond what is necessary for the object in view, namely, that my readers may learn very shortly the easiest routes by which to get to them; the kind of accommodation which they may expect on their arrival; the weather which they must be prepared for, so far as my own experience goes, and so far as I could ascertain from inquiries on the spot; and last, but by no means least, the occupations and amusements which the several places afford.

I wish to make my book as short as possible, in the hope that it may be of some use to invalids who, being advised to winter abroad, have great difficulty in deciding whither to go. These persons, in the great majority of cases, hear of some place, perhaps one of the well-known sanatoria on the Riviera, and buy a book on the subject, in which they read such delightful descriptions of the country, the beauty of the scenery, the magnificence of the climate, and the excellence of the hotels, that they forthwith decide that the place is exactly suitable to their wants; and they too often go there without further inquiry, furnished with a supply of light summer clothing, to find their expectations grievously disappointed, and to sigh in vain for the well-built houses, warm fires, good food, and other comforts of their English homes. I have had rather unusual opportunities of visiting many of the less well-known health-resorts; and it is with the view of giving my readers a chance of comparing the different places which I have visited, one with another, that I now sit down to write a short account of my travels in search of health, and I can promise that such account will be faithful to the best of my ability.

I shall carefully distinguish what I have learnt from my own observations, from what I have been told by others. I have no prejudice in favour of one place over another; and though I naturally preferred some of them to others, I shall give here the reasons for my preference, in order that my readers may judge for themselves whether they would be likely to be similarly influenced.

I venture to think, too, that a book of this kind may be of some use to those who have to advise their patients about winter quarters. It is impossible that doctors, with their numerous avocations, can wade through lengthy books of travel in different countries in order to gather the small details which they require in advising a patient whither to go, so as to secure the greatest amount of comfort with the least amount of fatigue and discomfort. Moreover, they know too well that many of these details, when found, cannot be relied on. I do not suggest that people knowingly write false accounts of places in which they have an interest, for the purpose of inducing others to go and stay there; but no one would believe, unless he had personally experienced it, how hard it is to get reliable information about a place from anyone who has lived there for some time, or who has frequently revisited it and formed a decided liking for it; such person has found the place suited to his own ailments, and he quite unknowingly exaggerates the advantages of it. Now I have no temptation to do this. I have changed my quarters nearly every winter, and I have thus been enabled to see what appear to me the advantages and disadvantages of each place from a health-giving point of view, and I shall not hesitate to state them.

I shall take the different places in the order in

which I visited them, commencing with the long sea voyage in a sailing ship to Australia. I am afraid that I must necessarily speak more about myself than is by any means agreeable to me; but this book, if of any value at all, will only be valuable in so far as it is a faithful record of my own personal experiences.

The great lesson which every invalid must learn, who means to profit by wintering abroad, is that he will have to exercise quite as much care and caution to avoid doing imprudent things out of England, as he has to do when at home; and, indeed, it is not too much to say that he will often have to exercise more, inasmuch as the temptations to neglect precautions are greater. If this be done, and winter quarters suitable to the particular cases be chosen, I have great faith in the efficacy of the remedy, when it is taken in time. I can only hope that many of my readers may as greatly benefit by winters abroad as I have myself benefited by them.

CONTENTS.

CHAPTER I.

THE LONG SEA VOYAGE TO AUSTRALIA.

Preparations for the Voyage.—Fares.—The Start.—Ship Surgeons.—Changes of Temperature.—Arrival at Melbourne.—Effect of Voyage on Invalid Passengers 1

CHAPTER II.

MELBOURNE.

Hotels.—Melbourne Club.—Colonial Hospitality.—Weather.—Dust.—Excursions in Victoria.—Big Gum Trees.—Native Reserves.—Bad Roads.—Suburbs of Melbourne.—Climate for Invalids 13

CHAPTER III.

TASMANIA.

Intercolonial Traffic.—Launceston.—Hobart Town.—Scenery and Vegetation.—English Fruits.—Mountains and Rivers.—Trout and Salmon.—Excursions.—Picnics.—Drainage of Hobart Town.—Sport.—Climate 26

CHAPTER IV.

SYDNEY.

Sydney Heads.—Port Jackson.—Beauty of Town and Suburbs.
—Hotels.—Clubs.—Woolongong.—Bulli Pass.—Cattle and
Sheep Breeding.—The Blue Mountains.—Bathurst.—Goulburn.—Australian Scenery.—Educational Advantages.—
Hot Winds.—Climate 38

CHAPTER V.

QUEENSLAND.

Brisbane.—Maryborough.—Sugar-Making and Refining.—Coolie
Labour.—Horse-Racing.—Australian Horses.—The Darling
Downs.—Sport.—Snakes.—Sheep-shearing.—Climate . . 50

CHAPTER VI.

THE RIVERINA.

Name and Situation.—Journey from Melbourne.—Bush Scenery.
—Station Life.—Salt Bush.—" Nipping and Shouting."—
Bush Inns.—Wild Turkeys and other Game.—Kangaroos.—
Scarcity of Water.—Horse-breeding.—Travelling in the
Bush.—Wagga Wagga.—Races.—Visiting.—Keeping Christmas.—Great Heat.—Climate.—Homeward Bound.—Australia for Invalids 69

CHAPTER VII.

THE VOYAGE HOME.

P. and O. Steamers.—Overcrowding.—Suez.—Alexandria.—
Valetta.—Climate of Malta.—Society.—Sport.—Routes
from England.—Home Again.—Results of Voyage . . 92

CHAPTER VIII.

ALGIERS.

Route from England.—" The Pearl set in Emeralds."—Hotels. —Wine.—Weather.—Mustapha Supérieur.—Villas.—"Zammitt's."—Climate.—Bad for Bilious Persons.—Drainage.— Want of good Hotel.—Marseilles.—The Bise and Mistral.— Pau.—Biarritz.—Arcachon.—Results of Winter . . . 100

CHAPTER IX.

EGYPT.

Routes.—Landing at Suez.—Cairo.—Shepherd's Hotel.—Climate of Cairo.—Dust.—Bad for Hemorrhage.—The Nile.—Drawbacks to Voyage.—Hiring of Dahabeah and Arrangements with Dragoman.—The " Gazelle."—The Nile Voyage.— Upper Egypt and Nubia.—Sport.—The North Wind.— Accidents to Boats.—Dragomans.—Helouan les Bains.— Luxor.—Cost of Living at Hotels and Expense of Nile Voyage.—Result of Trip.—Sicily.—Syracuse.—Palermo.— Home through Italy 119

CHAPTER X.

CAPE OF GOOD HOPE.

St. Leonards.—Voyage to the Cape.—Cape Town.—Hotels.— Drainage and Water Supply.—The " Cape Doctor."— Wynberg.—" Cogills " and " Rathfelders."—Scenery.—Constantia.—Vines and Wine Making.—Exorbitant Duties.— The Erste River.—Ostriches.—Railways.—Worcester.— Sport.—Weather.—Climate for Invalids.—Bloemfontein.— Methods of getting to it.—Ox Waggon and Gear.—" Trekking."—Expense of and Requisites for Journey.—Results . 146

CHAPTER XI.

DAVOS.

Change of Treatment in Lung Complaints.—The "Germ Theory."—Necessity for competent Medical Evidence.—Earlier Accounts of Place.—Situation and Surroundings.—Comparison with other Sanatoria.—Climate.—Cold and Dryness.—Toboggining.—Skating.—Weather.—Drainage.—Clothing.—Hotels.—Food and Wine.—Review of earlier Accounts.—Effect of Climate on Invalids.—Whether Lasting or otherwise.—Causes of Favourable Results.—The Snow-melting in Spring.—Where to go when it Commences.—Route from England.—Conclusion . . . 176

WINTERS ABROAD.

CHAPTER I.

THE LONG SEA VOYAGE TO AUSTRALIA.

Preparations for the Voyage.—Fares.—The Start.—Ship Surgeons.—Changes of Temperature.—Arrival at Melbourne.—Effect of Voyage on Invalid Passengers.

In the year 1872 I was recommended by my doctor to take a voyage to Australia on account of threatening mischief in the lungs. I was advised to take the long sea route, and I accordingly looked up all the notices in the newspapers of ships sailing for Australia, and was somewhat bewildered by the number of splendid clipper ships of different lines advertised to sail for different Australian ports; all offering first-class accommodation for passengers; and most of them promising the advantages of an experienced surgeon and a cow. I finally decided on trying Messrs. Green's well-known Blackwall Line, and I had no reason to be dissatisfied with my choice. The ship in which we sailed was roomy and fairly well found in provisions; and the officers, with the single exception of the experienced surgeon, to whom, in the interests of invalids, I think it my duty to refer hereafter, appeared to be competent in the performance of their duties, and were attentive to the wishes of the passengers.

The ship was advertised to leave the London Docks on the 5th of September, and to embark passengers at Gravesend on the following day.

A brother and cousin of mine, who were both in good health, had arranged to accompany me, and we paid a visit to the ship in dock to see what would be necessary to be done to make our cabins comfortable for the voyage. The cabins were large and airy, with good square windows, instead of the ordinary portholes, and each contained two good-sized beds or bunks, and a fixed washing stand and looking-glass; but there were no mattresses, nor was linen of any kind provided, and there was a general look of bareness in the cabins; which, as they were to be our homes for three long months, we considered ought to be remedied. I think that most persons would find that, in order to be tolerably comfortable, they would have to purchase bedding and linen, which however are sometimes provided by the shipowners, a carpet, curtains, a deck-chair and common chair for the cabin, and swing-tray. A chest of drawers, with a ledge round the top to form a dressing-table, is a great convenience, as are also canvas pockets to be hung round the cabin, in which can be kept all kinds of articles which are in daily use. All these things, and many other small articles which can hardly be called necessaries, but which add greatly to one's comfort on the voyage, and which cost but little, can be obtained at any of the outfitting establishments in the neighbourhood of the docks. We found that it cost about £13 each to furnish our cabins comfortably; and this sum, or something approaching to it, must be considered as part of the passage-money in estimating the cost of the journey, as the articles are of very little value at the end of the voyage. We had each of us a cabin to himself, the price of which was between £75 and £80; but two friends or relatives could quite comfortably join at one cabin, at a cost of about £55 each.

The only preparations in the way of personal clothing which are requisite for the voyage are plenty of warm under and upper clothing, with good thick boots or shoes, as the decks are often wet for many days together; a light woollen suit for the tropics, and one or two pairs of canvas deck shoes for fine weather. No proper washing of linen can be done on board, so that the quantity taken must be in proportion to the length of the proposed voyage. Paper collars, and several dozen cheap pocket-handkerchiefs, which can be thrown away when used, are very useful. In short, the preparations for a voyage round the Cape to Australia should be made much more in view of cold rough weather than of hot and fine: though in the tropics, where a sailing ship is sometimes detained by calms, the lightest clothing often feels almost unbearable. A good supply of light literature is essential.

We embarked at Gravesend at noon on the 6th of September, and anchored for the night off the Nore, and started again next morning at 7 o'clock. I had noticed on the first evening an oldish gentleman who sat next me at dinner as being very ill, apparently far gone in consumption : there were several other gentlemen travelling for health, but my next neighbour seemed very far the worst, and to be in need of very great care and attention. On enquiry, I found to my great surprise, that he was the experienced surgeon who was to look after us invalids. He got gradually worse, until, at his earnest entreaty, backed up by a written request signed by all the first-class passengers, the captain consented to put in at Falmouth and land him. We all felt sure that he could not have lived till the end of the voyage if he had remained on board; and it was evident to the most casual observer that he was totally unfitted to take charge of other invalids. On our arrival at Falmouth the captain telegraphed to the owners informing them of what he had done, and requesting that another surgeon might be sent immediately; and in the course of two days the new surgeon arrived, and we set sail again on our voyage.

The first day at dinner we discovered that our new surgeon was far from sober; and his intemperance grew daily worse and worse, and the only time that he was seriously wanted to attend some of the sailors, who had been injured in a storm, he was perfectly incompetent through drink to do anything whatever for them. Most fortunately none of the invalid passengers required much medical assistance, and no illness broke out amongst either the first or second-class passengers, otherwise I do not know what we should have done, as no one felt the slightest confidence in the surgeon or would have liked to trust himself to his care.

I have thought it right to state these facts, as it seems to me incumbent on ship-owners, who invite invalids to travel in their ships by means of advertisements of experienced surgeons being carried, to take care that such surgeons are not incapacitated from performing their duties by physical or moral disqualifications. I am bound to say the other advertised comfort, the cow, performed her duties to our entire satisfaction.

We had calms and contrary winds in the Channel, by reason of which, and of the delay caused by the change of surgeons, we did not leave Falmouth till the afternoon of the 17th of September, and on the same night we saw the light at the Lizard, and took our farewell glimpse of the English coast. We had fair weather through the Bay of Biscay, and on the third morning after our start we felt the first balmy feeling in the air, and were able to discard some of our warm wraps and great coats. We had then a few days of warm and pleasant weather, becoming gradually warmer till we got into the tropics. Then followed about three weeks of very trying weather: the heat was so great that we could hardly sit on deck during the day time even under the awning, and the saloon and cabins below became unbearably close and stuffy. The average range of the thermometer in my cabin during this period was from 78 to 84 degrees Fahrenheit, and the atmosphere was so charged with moisture that the effect on every one on board was that of being in a perpetual vapour-bath. It was difficult to employ ourselves in the day-time, and almost impossible to sleep at night. We lay with doors and windows wide open, but we could not get the least breath of fresh air, and the perspiration ran in streams off our hands when we put them out of bed. There was no refreshment even in our morning bath, as the temperature of the water was almost as high as that of the atmosphere, and the exertion of dressing left us in a complete state of exhaustion for the rest of the morning. I need hardly add that we had but little appetite for our meals. It was a trying time enough for the strong, but especially trying for invalids. It often happens that persons feel weak and feverish on first getting into the tropics, but this is greatly relieved by slight doses of quinine, of which invalids should be careful to take a good supply. This feverish feeling is generally followed by a rash, which, commencing on the neck, or on the hands and feet, gradually extends over the whole body. It is very irritating and troublesome, but is considered to be a sign that the fever is coming to the surface, and to be a matter of congratulation to the sufferers.

Between lat. 15° N. and the Equator, we had the usual light winds and heavy showers of rain, but fortunately we had no long continued calms about the

Equator. We crossed the Line about midnight on the 14th of October with a fresh S.E. trade wind, and then had a succession of very beautiful days till the 25th, when we got into latitude 28° to 29° south. These were the most enjoyable days during the whole voyage. The sun was very bright and hot, but there was always a pleasant breeze blowing, and we played games on deck during the day; and it was not too hot to sleep with comfort during the night. This fresh wind cannot, however, always be counted on so near to the Equator, in which case the traveller must be prepared for a longer spell of the hot weather with its attendant discomforts.

On the 25th of October we met with the first cold wind from the south, which wind continued, with very few days' intermission, all the rest of the voyage. We had two or three single days of warm weather, when there was almost a calm, after this date; but, roughly speaking, the cold winds continued and increased in severity till we sighted the Australian coast. During the latter part of our voyage there were days when the decks were quite slippery with ice, and a strict lookout was kept for icebergs. We could not sit on deck with any comfort, though wrapped up in shawls and great coats, and we even sat in our great coats in the saloon, and used them as additional coverings in bed, without succeeding in keeping ourselves warm. We had only a few days of really bad weather during the whole voyage, but there were, on the other hand, not a great many days when we did not suffer more or less from cold or heat; and we had certainly quite an average passage, at about the best time of the year.

The temperature in the southern hemisphere varies greatly from that in the northern. It is very much colder in the same degree of latitude south than it is in that degree north. For instance, London is in the 52d degree of latitude north, but in the same degree south there are icebergs throughout the year, and life would be insupportable to an European. Melbourne, the capital of Victoria, is in latitude 37° south, which is about the latitude of Malaga in the northern hemisphere, and yet it has, on the whole, a perfectly temperate climate all the year round—whilst Hobart Town, the capital of Tasmania, is in latitude 44° south, which is about the latitude of Nice in the northern hemisphere; and yet the climate is only like a very fine English climate throughout the whole year. This is attributable to the fact that in the southern hemisphere an almost unexplored icebound region extends to within about 65° of the Equator. In spite then of the fact that we never got into lower latitudes than about 47° south, we suffered a good deal from the cold, though it was then the commencement of the summer. When it is remembered that during the greater part of our time in the tropics it was almost impossible to move about on the deck during the day or to sleep at night, even without any bed clothing whatever, on account of the extreme heat, it will be seen how sudden and great was the change. I must at the same time say that I never caught cold during the voyage, though on one occasion I had a slight congestion of the lungs brought on by being on deck in the cold wind. I was, however, very careful; and it was only on one or two days in the tropics that I was on deck after sunset. Some of the other passengers had colds; and one of the invalids had an attack of pleurisy, which confined him to bed for two or three days: but in other respects the health of all the passengers was good; and the wonder is how few colds are caught on board ship, where every condition seems favourable to catching them—when caught, however, they are very difficult to get rid of. The food was fairly good throughout the voyage. We had sheep, pigs, and poultry on board, and we were never without a sufficient quantity of fresh meat. The cooking was fair, but the attendance both at table and in the cabins left much to be desired. We had our own supply of wine; but the spirits, beer, and porter on board the ship were very good. The captain was most courteous and attentive to the comforts of the passengers, and we had full confidence in the capability and discretion of all the officers who attended to the navigation of the ship.

We had one heavy storm which lasted about forty-eight hours, during which all the deadlights were up and the hatches shut down, and we had to move about as best we could in the dark below, or remain in our bunks. This was a miserable time; and one which no one can fully appreciate who has not had a similar experience. We were all more or less bruised by tumbling about in the dark, consequent on the eccentric movements of the ship; and three of the sailors were seriously injured by a wave which carried away some of the pig and hen coops with all the live stock in them. On the whole, however, we had a fair voyage, though it extended over more than the average length of time, which is usually estimated to occupy from eighty to eighty-five days. We caught our first glimpse of the Australian Coast on the 11th of December, the principal headland being Cape Nelson; but, owing to light winds, we did not make the Port Philip Heads till the 13th, and we anchored about a mile off Sandridge Pier at 10 o'clock the same evening. The coast line between Cape Nelson and the Heads is not particularly fine, consisting as it does of large tracts of sandy country covered with brush and scrub, and intersected by rocky headlands; but we were all extremely glad to greet it, and to see the near approach to the termination of our voyage.

It is often said that long voyages are so agreeable that the passengers are generally very sorry to get to the end of them, and to take leave of one another. I cannot say that this was our experience. We had no unpleasant bickerings amongst the passengers, and we got on fairly well together, but I think that all were very glad at the prospect of getting to the end of their journey, which had occupied ninety-nine days from port to port.

We were all weighed a day or two before we landed, and I was greatly surprised and disappointed to find that I had lost a stone and a half during the voyage, although I had kept in good health and had a good appetite the whole time. My brother had lost about half a stone; whilst my cousin had gained above a stone. I think that nearly all the invalids except myself had gained some addition in weight, and some of them a considerable addition. I felt stronger in general health than when I started, and my cough had slightly improved during the voyage; but it returned again when I got to land. I have no other means of judging whether or not my lungs were really benefited by the voyage; but the doctor whom I consulted in Melbourne on my arrival seemed to think, from what I told him, that I had gained some benefit, though not as much as he would have expected. I began to regain my weight soon after I landed; and I am inclined to think that life on shore suits me better than life on board ship, and I have now had considerable experience of both. Whether this is likely to be the case with others I cannot of course say. Some people, who are really fond of the sea and ship life, tell me that they are never so well as they are on board ship; that they eat and drink twice as much, and that their food seems to do them twice as much good as when they are on shore, and that they gain a steady increase in weight. I think that, if a long cruise could be made in a very comfortable and well-found ship, in moderate latitudes, say between 40° and 20° north, and with pleasant companions, it would do most invalids, whatever might be the nature of their disorders, a great deal of good; but when a voyage has to be made in a moderate ship, in all kinds of weather, and with such excessive variations of temperature as are entailed by a voyage from England to Australia round the Cape, I think that doctors ought to consider very seriously before advising their patients to undertake it.

CHAPTER II.

MELBOURNE.

Hotels.—Melbourne Club.—Colonial Hospitality.—Weather.—Dust.—Excursions in Victoria.—Big Gum Trees.—Native Reserves.—Bad Roads.—Suburbs of Melbourne.—Climate for Invalids.

I AM not going to describe the city of Melbourne, which has been sufficiently done in Australian books of travel. I may, however, say shortly that it is a fine, but not very beautiful town, as compared with the principal towns in the northern hemisphere. The streets are wide, and there are some very good shops and houses facing them; but they are often side by side with one-storied shanties, greatly spoiling the effect. The town has a new and somewhat unfinished look about it, which will of course improve year by year. The public buildings, including the Parliament House, Public Library, and Museum, and the Post Office, are particularly fine and admirably adapted for the several purposes for which they have been built, and there is a general air of business and prosperity about the place and people. The hotels are fairly good.

We went to Menzie's Hotel, which was comfortable, though, like all Australian hotels, rather too public and noisy for great invalids. There are two other good hotels in the place, called "Scott's" and "The Port Philip Club," of which I have no personal experience. I am told, however, that all the three hotels are about equally good, Menzie's being slightly the most expensive, but having decidedly the best and quietest situation.

After being there about a week we were made honorary members of the Club, and we were allowed to live there during the rest of our stay in Melbourne.

The Melbourne Club is one of the best clubs out of London, and we were thoroughly comfortable there, whilst we had the advantage of seeing and becoming acquainted with many of the gentlemen of the place, as well as with many of the squatters in different parts of the Colony, who all stay at the Club, when in Melbourne. There are clubs in all the principal towns in the Australian Colonies; and they are a very great advantage to travellers. They are all well managed and thoroughly comfortable; though they vary in style and comfort in accordance with the place where they are situated. The rules of these clubs admit of any gentleman who is visiting the Colonies, and who is properly introduced by a member, being at once made an honorary member, and as such becoming entitled to a bedroom and to live entirely at the Club. The expense is not more than at the hotels; and the comfort is incomparably greater. The custom is that an honorary member is admitted free of entrance money, and free of any subscription for the first month : after which he pays £1 per month subscription for as long as he remains. I believe there is a rule at all the clubs that no member, whether ordinary or honorary, can keep his bedroom for more than a month at a time without a break; but this seems to be very seldom acted upon, unless there happens to be a great run on the rooms at particular periods of the year, like the race-weeks. We were never asked to give up our rooms, though I stayed at the Australian Club in Sydney for nearly four months.

Throughout all the Colonies we met with the greatest kindness, consideration, and hospitality from every one with whom we came in contact. I believe that this is the experience of all decently conducted English travellers who visit the Colonies, in search of health or pleasure.

I am sorry to say that there are some of our fellow-countrymen who have taken advantage of this; and the natural consequence is that the residents are becoming more careful as to whom they invite to their houses; but no real English gentleman need fear meeting with the greatest hospitality throughout all the Colonies of Australia, especially if he have one or two good letters of introduction. There is no advantage in having many of these letters, but it is important that they should be addressed to the right people. We only remained in Melbourne for one month on our arrival, namely, from the 14th of December, 1872, to the 15th of January, 1873, and I was there again for a week just before returning home, namely from the

20th to the 26th of February, 1874, so that my knowledge of the place is somewhat limited. I cannot, however, speak favourably of the climate for consumptive invalids. No doubt the three months of December, January, and February are amongst the worst in the year in Melbourne, as that period is the height of summer, and the thermometer often ranges between 98° and 106° Faht. in the shade, whilst the hot winds frequently raise it several degrees higher.

When we landed, however, on the 14th of December, the weather was for the most part cold and wet, and remained so till the end of the month. We had a fire in the smoking-room at the Club on Christmas Eve, which was an almost unprecedented occurrence. The air was irritating to the throat and bronchial tubes, whilst the dust was almost unbearable even in that far from dry season. The streets being very broad and quite straight, the wind sweeps up the dust in such clouds that on several occasions, when standing on the steps of the Club which is at the top of Collins Street, the principal street of the town, I had great difficulty in distinguishing the lines of the houses on each side of the street at a little distance from where I stood.

I believe that the dust nuisance was unusually great at that time, as there were differences between the town and water authorities, which caused the latter to cease watering the streets. Under the most favourable circumstances, however, the quantity of dust must always be great in any wind, and highly prejudicial to persons suffering from complaints of the chest, throat, and respiratory organs.

The residents in Melbourne say that the spring and autumn are generally very pleasant; but other persons, who know the place well, have told me that the climate is very uncertain at all times of the year, and the air peculiarly irritating. On the whole, forming the best opinion I can of Melbourne from my own short experience and from what I learnt from others, I should say that it is not a place suitable as a residence for invalids with affections of the chest and lungs at any time of the year, though no doubt certain months are much worse than others.

There are some exceedingly pretty suburbs around Melbourne, such as Toorak, Richmond, and St. Kilda, all within easy distance of the town: and I should strongly recommend any invalid, who, from circumstances, is obliged to remain in or near Melbourne, to try and get accommodation at one or other of these places. I have heard of boarding-houses in these suburbs, where visitors are made fairly comfortable at reasonable prices, but I have no personal experience of any of them.

Interesting excursions can easily be made from Melbourne to Geelong and Ballarat, and further into the western part of Victoria, where are some of the finest sheep and cattle stations in Australia. Ballarat was the centre of the alluvial gold fields in the early days of the Colony, and it still preserves some of its former grandeur in its fine large banks and public buildings; but the alluvial gold fields are now almost worked out, and the attention of gold miners is turned to the crushing of quartz in other districts of the Colony. Ballarat is now a quiet and pretty country town, and is said to be a particularly healthy one. From its situation I should think that this is the case, but it would be a dull town for an idle man to reside in for any length of time.

It is also an easy journey from Melbourne by sea to Gippsland, where is some of the most beautiful scenery in the Colony with a fine climate ; but invalids should not attempt this unless they have letters of introduction to some of the squatters who are resident there, as they could get no proper accommodation in the district.

There are numerous other excursions to be made from Melbourne by Cobb's coaches, or in a private buggy, which are well worth making; but full information should first be obtained as to the nature of the roads to be traversed and the accommodation which can be obtained. We made a very beautiful excursion for five or six days in a buggy to what is called " The black spur range of the Dandenong Mountains," which lie between Gippsland and the rest of the Colony of Victoria; the road runs for miles through magnificent primeval forests of the native trees of Australia. Here are the big gum trees, not perhaps comparable in individual height to the big trees of California, but more remarkable in that they form mighty forests which extend for miles and miles in every direction, the trees being of an average height of nearly 200 feet each, with a proportionate bulk. We saw one tree which had just been cut down, and the trunk of which measured 276 feet in height, with a girth of thirty-six feet in circumference at the thickest part. The woodman who had cut it down told us that it was by no means an unusual size, and that he had cut one down in the same forest which measured over 420 feet in height with a proportionate girth, and which had been sent as a specimen to the English Exhibition of 1851. It was a feature in these large trees that they threw out no branches for the first fifty or sixty feet, and it was consequently possible to see for a great distance around us when we stood in certain parts of the forest.

The leaves of gum trees are so arranged that they afford wonderfully little shade from the sun, and the rays streamed down through the trees on all sides, lighting up the beautiful undergrowth of ferns and mosses and parasitic plants, in a manner which it is quite impossible adequately to describe. The intense stillness was only broken by the screams of numberless parrots of different species, but all of gorgeous plumage, which fluttered about from tree to tree, and by the occasional cry of the Australian pheasant, or Lyre bird, which is still found in that district, although the demand for its beautiful feathers has greatly reduced the numbers, and made the bird exceedingly shy and difficult to approach.

On this expedition we visited a place called Healesville, where the Colonial Government has established one of its reserves for the small remnant of the aboriginal inhabitants which is still left in the Colony.

There are between 400 and 500 of these natives gathered together and living in huts round the house of the Government Superintendent. There is a school for the children, and every effort is made to teach both young and old the way to get their living by husbandry and simple handicraft. From what I could hear, however, the natives do not seem to take to civilised ways, and it seems probable that they will soon become extinct, except in those parts of Australia which are still uncolonised by white men. There are fair country inns at convenient halting places for this expedition, but it is hardly one that can be made by great invalids. The roads are in some places fearful, not unlike a heavily crevassed glacier, with the difference that the crevasses run in the same direction as the road instead of across it. These crevasses are made by the heavy bullock waggons in wet weather, and they are often from one to two feet in breadth and four or five in depth. It is no uncommon thing for passengers in the public coaches to be seriously injured by the terrific jolting over these so-called roads.

I am informed that since we were in Australia a railway has been made between Melbourne and Sale, the principal town of Gippsland; and it is probable that much of this beautiful scenery can now be seen with very little trouble or inconvenience. The roads in the immediate neighbourhood of Melbourne are very good; and interesting walks, rides, and drives can be taken without fatigue.

The river Yarra Yarra is an exceedingly pretty

river; it is narrow and overshadowed on both sides by beautiful trees and shrubs; and it is a good and safe river for boating. There is an excellent cricket ground, where the game can be played on nearly every day throughout the year; and a fine race-course, with a grand-stand and all the other conveniences of a first-class race-course in England. There is good hunting during the winter months, and fair horses can be bought or hired at reasonable prices. There are numerous picnics, balls, and dinner-parties, to which English travellers, especially if staying at the Club, are hospitably invited upon very slight acquaintance; and society is not nearly so formal or formidable an undertaking as it is in England.

I think I have said enough to show that Melbourne is by no means wanting in objects of interest and amusement to those who are in fair health. I am far from wishing to say too much against it for invalids; but I am sure that they should exercise more care and caution whilst there than is necessary for them in most other parts of Australia. It is especially important for invalids who arrive in Melbourne after a sea voyage to be very careful in the matter of clothing whilst they remain there. The changes of temperature in a single hour are often extraordinary; a bright, still morning will frequently be succeeded by a cloudy and gusty afternoon, the wind being exceedingly cold and penetrating. I was caught in one of these sudden storms the second day after we arrived at Melbourne. The morning was fine and bright, and I had taken an umbrella to keep off the sun. Suddenly the sky clouded over and there was a heavy shower of cold rain, succeeded by high gusts of wind and clouds of dust. I had warm underclothing, and fortunately escaped without ill effects; but a new black coat which I had on was completely ruined by the rain and subsequent dust. This is such a well-known occurrence that nearly all the residents wear dust-coats to preserve their clothing, even when walking about the streets on business or pleasure; and dust protectors of all kinds seem to be one of the principal articles of commerce in the shops. Woollen clothing, light in colour and weight during the summer, and above all a flannel vest next the skin, are necessaries for health in Melbourne; as indeed they are in most other parts of the world.

I heard a good deal when in Melbourne about the badness of the drainage, and indeed of the total absence of any proper system of drainage whatever in some parts of the town. I cannot remember that we had any personal inconvenience from this cause; but it is almost inevitable that it should exist. The town is built on low swampy ground within a few miles of where the river Yarra Yarra flows into the Bay of Port Philip: some part of the town is below the level of the river; and in wet weather, one of the principal streets is often one or two feet under water. I am not aware that the town is specially unhealthy, or subject to severe visitations of the numerous diseases which generally attend on bad drainage; but these diseases no doubt exist in a greater or less degree.

I was told by a doctor in Melbourne that for many years after the settlement of the first colonists, diseases of the chest and lungs were almost unknown there, but that in later years they had greatly increased, and that they now form almost as large a proportion of the causes of death in Melbourne as they do in the large towns of England. My informant attributed this in a great measure to the large importation of English people who had gone to Melbourne with the seeds of those diseases in them, and who had been obliged to leave England on that account. He would not admit that the climate was bad in such cases; indeed, he said he knew many persons who had arrived at Melbourne in apparently a very advanced state of consumption, and who had rapidly rallied, and had gradually become able to live and carry on their work there in comfort; these persons married in course of time and left families, to whom, he said, they handed down their own consumptive tendencies, which thus spread and became epidemic in the place. At the same time the doctor admitted that he did not consider Melbourne by any means the best place in the Colonies for persons suffering from affections of the chest and lungs, and he strongly advised such persons to go, if possible, into the interior of Victoria or New South Wales, and well away from the sea coast.

I should recommend invalids who arrive in Melbourne during the months of November, December, or January to postpone seeing the place and neighbourhood till a more favourable time of the year, and to start as soon as they conveniently can do so after their arrival for Tasmania or New Zealand. We did not visit New Zealand, so I cannot speak from personal knowledge of the climate; but I have been told on very good authority that, during the summer months, the climate of the southern island is delightful, and the accommodation in the towns quite sufficiently good. The scenery is in some parts very grand, much grander than anything that is to be found in any of the Colonies of Australia.

During the month which we spent in Melbourne I regained nearly half a stone of the weight which I had lost on board ship; and in other respects I felt better and stronger at the end of our visit than when we landed. I was able to take fairly long walks on level ground without any great fatigue, which I had been wholly unable to do before leaving England. Two out of our three invalid fellow passengers who remained in or near Melbourne told me that they felt better and stronger for the change, but they intended to return to England by an early ship round Cape Horn, in accordance with the advice of the doctor in Melbourne whom they had consulted. The same doctor had advised me to adopt a similar course, as he had more faith in the value of a sea voyage than in that of a residence in the colonies. I had, however, had enough of sea voyages for a time, and I am glad to say that I did not follow his advice. The third of our fellow passengers, who seemed to be the greatest invalid of the three, did no good in Melbourne, and, having friends in other parts of Australia, he soon left to join them. I saw him about six months afterwards in Queensland, and he told me he was much better and stronger, and he intended to remain in Australia for a considerable time.

I made the acquaintance of several young Englishmen in the different colonies of Australia who were travelling for health, principally on account of delicate lungs; they were all in different stages of the disease, but I only know of one who seems to have made any real advance to restored health, and he spent the greater part of his time on a high-lying station in New South Wales, in a particularly healthy part of the country. Two others died in Melbourne and one in Queensland soon after we left the country. I think, however, that if proper precaution be taken to choose the healthiest districts and to live the most careful lives, a voyage to Australia would be productive of good to many persons affected with lung complaints. But I should not recommend a long residence in Melbourne. We remained there about a month, and then went on to Tasmania.

CHAPTER III.

TASMANIA.

Intercolonial traffic.—Launceston.—Hobart Town.—Scenery and Vegetation.—English Fruits.—Mountains and Rivers.—Trout and Salmon.—Excursions.—Picnics.—Drainage of Hobart Town.—Sport.—Climate.

TASMANIA, formerly called Van Diemen's Land, is an island lying due south of Victoria. The town of Launceston, on the north side of the island, is about twenty-four hours' distance by steamer from Melbourne, and communication is kept up between the two colonies by steamers of the Australian Steam Navigation Company, familiarly known in the colonies as the A. S. N. Co. Nearly all the passenger traffic between the several colonies is carried on by the steamers of this company, which are, on the whole, fairly good and well found in provisions; but often overcrowded with passengers, and thus rendered far from comfortable. We started in one of these steamers on the 15th of January, 1873, for Launceston, arriving there on the following day.

The sea is often rough in Bass's Straits, and, as the steamers are small, travellers must be prepared for a considerable amount of pitching and rolling. We had our share of both.

Launceston is prettily situated at the mouth of the River Tamar, but is otherwise an uninteresting town; and everything that is worth seeing there can be easily seen in one or two days. There are two or three fair inns in the town, but nothing to tempt a visitor to stay longer than is necessary to see the place. When we were in Tasmania a coach ran daily from Launceston to Hobart Town, which is on the south side of the island, and which is the place where most visitors in Tasmania prefer to reside. The coach started at five o'clock in the morning, and arrived at Hobart Town at eight o'clock in the evening, thus making a long journey of about fifteen hours; but the journey might comfortably be broken by staying at a place called Campbell Town, between five and six hours distant from Launceston, where there are two or three decent inns. The road passes through the middle of the island, and is good all the way; and the scenery is exceedingly pretty, without being grand. Since we were in Tasmania a line of railway has been opened between the two towns of Launceston and Hobart Town; and no doubt the journey can now be made easily in six or seven hours.

There are several inns at Hobart Town, but I cannot greatly recommend any of them. In the Australian colonies all the inns are called hotels, however poor the accommodation may be; and travellers in the country districts must not be surprised if they commonly find little but the name to remind them of the comforts of European hotels. There is generally plenty to eat and drink, but things seem sometimes rather rough to people who have not lived all their lives in the bush. This is even the case with the hotels in Hobart Town. There is a small club in Hobart Town which has been lately built, but there are as yet no bedrooms in it, which a traveller can occupy, as in the other Australian capital towns. The best course for a visitor to take, if he be not satisfied with his hotel, is to look out for private apartments, of which there is a fair supply in the town. The custom there is for the owner of the house to provide his lodgers with board and lodging at a fixed weekly price. The terms are for the most part reasonable, and, so far as our experience went, the accommodation was very good, and the food excellent and abundant.

Hobart Town is the chief town of Tasmania, and contains about 20,000 inhabitants. It is an exceedingly pleasant town, and is beautifully situated at the mouth of the river Derwent and under the shade of Mount Wellington. The town has not an imposing appearance, as there are but few large buildings in it, and most of the houses in the streets are rather low. It looks like an English cathedral town in its quietness and comparative antiquity. Indeed, the principal thing which strikes the English traveller in Tasmania is its singular resemblance to his own country. The very names are all English, perhaps too much so. He sails up the River Tamar, and lands at Launceston; he crosses the island to Hobart Town, and finds the River Derwent, with Mounts Wellington and Nelson overhanging the town. The climate is similar to that of England, only very much finer and brighter, and nearly all English flowers and fruits grow there in the greatest abundance, and of most excellent quality. The trees and general vegetation are of course very different, but not so much so as to dispel the illusion which anyone may readily fall into in the neighbourhood of Hobart Town that he is walking about in one of the prettiest of the southern counties of England. There are English-looking houses surrounded by English-looking gardens in the suburbs of the town; and beyond them large fields of hops, with drying kilns erected on them, very similar to what we see at home.

Tasmania is each year becoming a larger hop-growing country, and Tasmanian beer is extensively drunk in the Colonies, and is very good and much lighter than English beer, which has to be brewed of extra strength for exportation to the Colonies. Moreover the general features of the country have a far greater resemblance to England than they have to the mainland of Australia. The island is small, and everything is on a comparatively small scale; but there are mountains and rivers to be seen in all directions, which are both conspicuously wanting on the mainland, and the eye roams over a spacious park-like landscape. The vegetation is most luxuriant by reason of the quantity of water which is to be found everywhere; and English fruit trees attain a size and quality which they hardly ever acquire in their native soil. The peaches, pears, apples, currants, and gooseberries are all most excellent, and the mulberries such as I never saw before or expect to see again.

Hobart Town is the great emporium for jams and preserved fruits for the other Australian Colonies. It is also the pleasant and health-giving resort for such of the inhabitants of those Colonies as can spare the time and money to spend there the hottest part of their own summer season at home.

Tasmania has in late years become a great field for mining enterprise. The tin mines are extraordinarily rich and easy to work; and gold, silver, iron, and coal are found in very paying quantities. The population of the island is small as compared with its size, and there is a great want of steady English labour to assist in developing the natural resources of the island.

We remained in Hobart Town till the 8th of April, and enjoyed our visit exceedingly. There again we met with very great kindness and hospitality, though, from motives of prudence, I went out very rarely to dinners or evening parties. This is a rule which I am sure ought to be observed by invalids in Australia quite as much as in England, as the changes of temperature between the days and nights are even greater than at home.

During the time that we were in Hobart Town the weather was, for the most part, clear, bright, and sunny, and neither too hot nor too cold; but there is on many afternoons, a rather treacherous sea-breeze which springs up suddenly and causes a very sensible lowering of the temperature. The outward and visible sign of this sea-breeze is a covering of the upper part of Mount Wellington by a thick white cloud, which descends nearly half way down the mountain. When this comes on, it is time for the invalid to return home and remain indoors.

The walks, rides, and drives in the neighbourhood are beautiful; and those who are fond of boating and sea-fishing can get any quantity of both. Interesting excursions can be made to Port Arthur, the old convict station when England used to transport her criminals to Tasmania, and where the last remnants of the convicts who have not been liberated or who have been re-convicted in the Colony are still living; and, up the river Derwent to New Norfolk, in the immediate neighbourhood of which place are the salmon ponds, whither the ova of trout and salmon have been carried from England, and reared with great care, and with a considerable amount of success. The trout have grown and multiplied, so that some of the rivers in Tasmania already afford excellent sport to the fisherman. The salmon did not succeed so well at first. Indeed, when we were in Tasmania, it was greatly doubted whether any true salmon had survived its birth, and was then in existence in Tasmanian waters. A reward had been offered by the Government to the first person who caught one, and though the reward had been claimed on one or two occasions, the claim had always been rejected on the ground that the fish caught were not true salmon. I tasted, however, one of the largest and best salmon-trout which I ever saw; it was caught in the Derwent whilst I was in Hobart Town, and was sent as a present to the governor: it was eight or nine pounds in weight, and was as pink in flesh and quite as good as any salmon could be.

I believe that since that time some true salmon have been caught in the Derwent, thus setting at rest a much vexed question in the Colony.

A charming excursion can be made by riding or driving to the river Huon, where there is a fair inn. The road passes through grand old forests of gum trees; and winds about amongst the wonderful fern tree gulleys which are amongst the chief and most beautiful characteristics of the island. The walks in the immediate neighbourhood of Hobart Town are numerous and well adapted to both strong and feeble pedestrians. Picnics are the chief amusements of the place during the summer; and, if proper regard be had to weather and clothing, are pleasant and healthful recreations; especially when it is remembered that the young ladies of Tasmania are renowned for being amongst the prettiest and most fascinating in the Colonies. The food is good and plentiful, and the prices seemed to be most reasonable for all necessaries of life. The mean temperature in the summer is about 63°, and in the winter about 46° Fahrt., so that there are no great extremes of heat or cold; and the annual rainfall averages about 24 inches. I was told that snow seldom lies in the valleys or plains, but the highest mountains are capped during two or three months in the winter. There are often sharp frosts in winter, but they are generally accompanied with clear bright sunshine in the day time.

From what I could learn from residents and visitors, I believe that an invalid might live comfortably and advantageously in Tasmania during the whole year; but I incline to think that, if possible, he should go to some warmer colony on the breaking up of the summer for two or three months, even if he were to return to Tasmania for the winter.

I cannot, however, speak from personal experience of the climate of Tasmania except during the summer months; as, although I paid a second visit to Hobart Town before leaving Australia, it was in the month of January. The pleasantest course to take, and which, I think, is the most healthful for an invalid visitor in Australia, is to move from colony to colony, gradually getting northwards as the winter progresses; and this was the plan which we adopted.

The drainage of Hobart Town left much to be desired. A large open main sewer ran through the centre of the town, the exhalations from which were at times powerful and noxious. The odours, too, in some of the smaller and less frequented streets were far from pleasant, and called to mind the probable existence of typhoid and other fevers in the place. We had a most grievous evidence that typhoid fever did in fact exist, and in a most virulent form. My cousin, a strong and healthy young man, was attacked by it within a few days of his arrival in Hobart Town, and he died in about three weeks, in spite of the efforts of the three principal doctors in the place to save him. It is not known exactly how or when he imbibed the germs of the fever; but there was an open sewer not far from the hotel where he was staying; or it is possible that he may have drunk water containing these germs. The fever was most acute and rapid, and there seemed to be but little hope of his recovery after the first few days. This melancholy event will always cast a deep shade of sadness over the recollection of some of the pleasantest days I spent in the Australian colonies.

The summer began to break up about the middle of March, and we had a fortnight of showery and rather cold weather; but it afterwards turned fine, though not so warm, again, and remained so until the end of our visit. I had no opportunity of staying at any place in the interior of the Island, but I incline to think that, if proper accommodation could be found, there are places which, from the absence of the sea breezes, would be preferable to Hobart Town, from a health-giving point of view.

There is fair sport to be obtained in most of the country districts of Tasmania, but it can only be properly enjoyed when visiting the stations of the squatters in the neighbourhood. Large herds of kangaroos are found in every direction, and there is capital amusement in hunting them with hounds which are bred for the purpose. There are still plenty of rabbits left in some parts of the Island, but they have been greatly thinned down since their skins became valuable for the making of felt hats and other articles of apparel. At one time they were a perfect pest not only in Tasmania but in many parts of Victoria. They were introduced into the Colonies many years ago, like the thistle, by some patriotic Englishman or Scotchman, and they both multiplied exceedingly in the fine climate of their new home. The thistle still flourishes and defies all efforts for its extirpation; but the rabbits have in late years been greatly reduced in numbers, though they have cost a great expenditure of time and money in the process.

I have had pointed out to me in Tasmania large tracts of good country which were at one time given up entirely to the rabbits, as it cost more money to keep them down than the land was worth; and it was only since they became worth destroying for their skins that the land has again been brought into cultivation. There are wild ducks of all kinds on the coast and on the rivers and inland creeks, and quail and snipe in the right season. There is no sport, however, to be got in the immediate neighbourhood of Hobart Town; and as we had unfortunately no introduction to any of the squatters in the island, I cannot speak from personal experience of the best districts for it.

I certainly improved considerably in health and strength during our three months at Hobart Town. I was able to take longer walks in the country, and I was not obliged to confine myself to the level roads. I walked to the top of Mount Nelson, which is about two thousand feet in height, and more than half way up Mount Wellington, which is over four thousand, before we left Tasmania; and I felt no ill-effects from the unusual exertion. I ate, drank, and slept well, but I regained very little of the weight which I had lost on board ship, either in Tasmania, or at any subsequent time during our stay in Australia.

The food in all the colonies is excellent. The beef and mutton are, I think, better than what we have in England. Both are for the most part grass-fed, and not so rich, though quite as nutritious as what we get at home—and every kind of meat is much cheaper and more abundant than in England. It is, moreover, cooked to suit English tastes; and the out-of-door life, which the fine climate in most of the colonies admits of, gives invalids every opportunity of getting a proper appetite to enjoy it. I can recommend nothing more conducive to health or pleasure than a long visit in Tasmania; and I should do this with the greater confidence if the person going there were provided with one or two good letters of introduction to squatters who have stations in the interior of the Island.

Invalids who require a warm, dry, and equable climate throughout the whole year would find it to a greater certainty on the mainland of Australia, where they can get far away from the disturbing influences of the sea; but English-born persons who are only in search of a finer climate than they can get at home, with something like the same variations of temperature in the different seasons, but with much more sun and much less rain and cloud, will find, I believe, nearly all they require in the beautiful Island of Tasmania.

CHAPTER IV.

SYDNEY.

Sydney Heads.—Port Jackson.—Beauty of Town and Suburbs.—Hotels.—Clubs.—Woolongong.—Bulli Pass.—Cattle and Sheep Breeding.—The Blue Mountains. — Bathurst. — Goulburn. — Australian Scenery.—Educational Advantages.—Hot Winds.—Climate.

WE left Hobart Town in a steamer called the *City of Hobart* on the 8th of April, 1873. This steamer did not belong to the A. S. N. Co., which has no service between Tasmania and Sydney : but the same remarks which I have made about the steamers of the A. S. N. Co. are applicable to the *City of Hobart*.

We had a fine passage, and on the morning of the 11th of April we came in sight of Sydney Heads, and entered the harbour about 10 o'clock. A thick, drizzling rain had come on, and we could not possibly have seen the beautiful harbour of Sydney under less favourable circumstances ; but even with these disadvantages we were greatly struck with the grandeur of the entrance, and the remarkable beauty of the bay. Sydney is most beautifully situated at the head of the harbour, which is commonly called Port Jackson, and immediately opposite to the entrance of the river Paramatta. The town is built on the side of a hill which slopes down to the water, and bears some resemblance in its position to the town of Naples. It is a much disputed point amongst travellers which of the three harbours, Port Jackson, the Bay of Naples, or the Harbour of Rio Janeiro, is the most beautiful. I have not seen Rio, but I give the preference to Port Jackson over the Bay of Naples from the singular variety of its scenery. The harbour is ten miles in length from Sydney to the Heads, and in no part more, I should think, than two miles in breadth. In two places the opposite shores are so far projected towards one another as to reduce the breadth to a little more than half a mile, and the harbour has thus the appearance of being divided into three separate parts connected by narrow inlets. The entrance at the Heads seems to be little more than sufficient to allow two ships to pass one another, though no doubt the breadth is much greater than it appears.

The shores are thickly wooded to the edge of the water on both sides, and the harbour is fringed with countless little bays separated from one another by narrow promontories, on which are built the picturesque houses of the principal inhabitants of the place, surrounded by bright flowering gardens and orange groves. On the nearest to Sydney of these promontories stands Government House, with its beautiful garden reaching down to the water; and nearly adjoining to it are the Botanical Gardens, full of luxuriant trees and shrubs, and blooming flowers, which extend for a considerable distance along the shore.

Very few of the more wealthy inhabitants of Sydney live in the town : some live in the adjoining suburb called by its old native name of Woolloomoolloo; whilst others have built charming houses and laid out beautiful gardens further and further from the town, until nearly the whole of the southern shore of the harbour is studded with them from Sydney to the Heads. There is also a beautiful suburb on the north shore which extends for some distance up the river Paramatta; and nearly all the houses are built in good taste and add to the beautiful scenery of the harbour : they are often covered with bougainvillias and bignonias, and other flowering creepers, whilst the gardens are bright with many of our choicest hot-house flowers. The Botanical Gardens at Sydney are beautifully laid out and well kept up, and they are celebrated for their great interest to the botanist as well as to the general visitor. And here I may say that in all the colonial capital towns, the places of public recreation and instruction are on a very much finer scale, and conducted on far more liberal principles, than they are in England.

The museums, public libraries, and public gardens, are open to all respectable persons at all reasonable hours of every day free of charge, and without any order being requisite; and in all the libraries are comfortable rooms for reading, and competent persons to find and bring the required books to the readers. There is also a circulating library, attached, I think, to all the public libraries, of books which may be taken home by any respectable resident or person introduced by him. The museums are well arranged, with the names and species of the different objects legibly written on them: and we always found the curators most willing to go round with us and give us any information we might require.

Sydney is a much older-looking town than Melbourne, and it is not laid out in the same formal manner. The principal street is long, and not very broad, and there are good shops facing it: but the public buildings are not nearly so fine as at Melbourne.

There are two or three large hotels in the place, but I can only speak from personal experience of Petty's Hotel, which is very comfortable and with very moderate charges. We went there on our arrival at Sydney and remained a month, when, having been made honorary members of the Australian Club, we took up our quarters there, and lived most comfortably during the remainder of our visit in Sydney. There is another club in Sydney called the Union, of which we were also made honorary members, and which seemed to be quite as comfortable as the Australian; and it has the advantage of having a small garden around it. There are numerous lodging and boarding-houses in the place, some of them very well situated, and which I heard well spoken of as clean and comfortable, and very reasonable in their charges; and these are better fitted than the hotels for invalids or families who may be making a long stay in Sydney.

The day we arrived at Sydney continued wet and rather cold, and we had unsettled weather for ten days afterwards; but, after the 20th of April, we had above a month of uninterrupted beautiful weather; and, indeed, during the whole of our stay in Sydney, till the 15th of July, we had very fine weather, broken by an occasional wet and cold day, but for the most part fine, dry, and warm, without being at all too hot. It will of course be remembered that the seasons in the southern hemisphere are exactly opposite to the seasons in the northern; and that therefore the period of our stay in Sydney comprised the latter part of the autumn and the early part of the winter.

The country round about Sydney is very pretty, and there are charming walks, adapted to good and indifferent walkers, and beautiful rides and drives in the immediate neighbourhood. Steamers ply daily in the harbour and up the river Paramatta, and excellent sea fishing may be got in nearly all the bays of the harbour. There are numerous excursions of two or three days which may be made from Sydney, and which well repay the traveller for his trouble. The scenery of the Hawkesbury River is considered to be as beautiful as anything in New South Wales; but as there is as yet no line of railway to the district, and no good accommodation to be met with when the traveller gets there, it is less known than it deserves to be, considering that it lies within forty miles of Sydney.

A pleasant excursion may be made to Woolongong, a town on the shore of the Pacific Ocean, a little to the south of Sydney; there is a railway to Campbell Town, and the rest of the journey may be performed on horseback or in a buggy, over a beautiful pass called the Bulli Pass. The vegetation on each side of this pass is most luxuriant. Magnificent palm and fern trees grow in profusion, in most graceful forms, and some to a considerable height: these are connected and interlaced with climbing plants of all descriptions, the branches of which, tinted when I saw them with every shade of autumnal colouring, hung in festoons from tree to tree, and created a succession of golden bowers on both sides of the road. The Bulli Pass is the only place in any of the southern colonies of Australia where the palm tree attains to any luxuriance in an uncultivated state; and it is therefore well worth a visit by those travellers who do not propose going as far north as Queensland. There is a fair inn at Woolongong, but it is not a place of much interest in itself. We had, however, a rather amusing introduction to it. I went there with the Governor of Tasmania, who was staying in Sydney, and who had kindly invited me to accompany him. When we got to within about a mile of the town, we saw what appeared to be two old English four-wheeled cabs on the road in front of us, the occupants of which were apparently picnicing on the side of the road. As soon as our carriage appeared in sight, the people hurried into their cabs and drove quickly back into the town. We took no particular notice of this occurrence, but we afterwards learnt that the inhabitants, being desirous of showing honour to the Governor of Tasmania, had sent out a deputation to present him with an address of welcome to their town. Whether their hearts failed them at the last moment, or what other cause prevailed, I know not; but certainly the address was never presented, and it was quite by accident that we afterwards learnt the facts.

On another occasion we went, by the kind invitation of a gentleman who is well known in New South Wales as one of the most successful breeders of shorthorned cattle in the colony, to spend a few days at his house on the river Nepean, about fifty miles by railway from Sydney. The scenery was extremely pretty; and this was our first introduction to that most delightful and health-giving existence called station life, which I shall attempt to describe in a subsequent chapter. I am no great judge of shorthorned cattle, but we saw some splendid looking animals on our friend's station, and such as would, I think, have been hard to beat in the most renowned herds in England. There was one bull for which our host had refused seventeen hundred pounds, and a heifer which had cost him over seven hundred; and he had no difficulty in selling his young two-year-old bulls for five hundred pounds each. So far as I could see, the breeding of cattle and sheep was conducted on much more scientific principles throughout all the Colonies of Australia than is commonly the case in England, and the result is that the herds and flocks have attained a high degree of excellence, and that there is now very little demand by the Australian squatters for the high-priced stock which they formerly imported in large quantities from England.

The principal excursion which every visitor makes from Sydney is by railway over the Blue Mountains to a town called Bathurst, which is a fair-sized country town at the terminus of the Great Western line of railway. The inhabitants of Sydney never weary of telling visitors that this line of railway is the finest specimen of engineering skill in the world—and it certainly is very fine. The line crosses the Nepean River at a place called Penrith, and immediately commences to ascend the Blue Mountains. The gradients are extraordinarily steep, and in one place the train has to go backwards and forwards along zig-zags in order to make the ascent.

The scenery through the Blue Mountain ranges is very beautiful, and no traveller should pass through them without spending a day or two at Mount Victoria, which is nearly at the summit. There is a fair inn there, and the views in the immediate neighbourhood are most beautiful. We spent three charming days there in the early spring, and the air was bright, clear, and exhilarating, and quite frosty at night. Mount Victoria is about 2000 feet above the sea level, and it would make an excellent site for a sanatorium or large hotel, if any person were bold enough to make the experiment of building one there.

We stayed about a week at Bathurst, on the other side of the mountains, and we found the air there also very pure and good, and much cooler than Sydney. I do not know the altitude of Bathurst, but I should think it must be 500 or 600 feet at least above the sea level. I did not visit Goulburn, which is at the terminus of another branch of the same railway, but it lies rather higher than Bathurst; and I was informed by an invalid friend, who spent many months there, that the climate was beautiful and cool, and that he benefited greatly by his residence there. Either of these towns would make an agreeable change from Sydney in the hot weather, and there are fair inns at both places.

Visitors in the Colonies of Australia must not expect very grand scenery in any of them. The country is for the most part flat or undulating, with no background of lofty mountains to relieve it; and there is a noticeable absence of rivers or running streams. The forest trees and shrubs and the fern trees are in some parts very beautiful; but the general tone of the country is of a somewhat monotonous olive green, contrasted, I can hardly say relieved, by darker coloured scrub in the bush districts. The wild flowers and flowering shrubs are in places very beautiful, and notably on the ranges of the blue mountains; and the general vegetation is so unlike what we have in Europe that it cannot fail to interest Europeans. There is a great scarcity of wild animals and birds, except in those districts where kangaroos abound; but those which the traveller sees on his journeys are strange and unfamiliar; and some of the birds, especially of the parrot tribe, are beautifully plumaged, though, for the most part, with very unmusical notes. Travellers who desire grand mountain scenery, with snowy peaks and glaciers, must go to New Zealand.

There is a large and very pleasant society in Sydney, and we met with the greatest kindness and hospitality from the residents during our visit there. There is plenty of work done both by the mercantile and professional men in the town; but the inhabitants of Sydney do not seem so thoroughly immersed in business as they do in Melbourne; and there were few days when we were not invited to join excursion or fishing parties in the harbour, or to spend the afternoon in some of the charming gardens along its beautiful shores; and there was very little fear that these outdoor amusements would be spoilt by the weather.

There are excellent Universities both at Sydney and Melbourne, with a competent staff of professors and lecturers at each, and where a good education is provided on very reasonable terms to the students. The university buildings are good in both towns, those in Melbourne being perhaps the finest; whilst Sydney has the advantage of having in Dr. Badham, the Principal of its university, one of the finest Greek scholars of his day, as well as a most accomplished gentleman and a very genial companion. There are good grammar and preparatory schools in both towns.

We spent a much longer time in Sydney than we did

in Melbourne and at a better time of the year, and we became more intimate with the families of many of the residents there; and it is possible that I may be somewhat prejudiced in favour of the place; but, making allowance for these favourable circumstances, I should decidedly recommend Sydney in preference to Melbourne to invalids who might have to pay a lengthened visit in one of those towns. I have grave doubts whether either place is particularly suitable for invalids who have any serious affection of the lungs; but in cases where persons having a consumptive tendency are obliged to reside out of England, and yet to live in a town in order to follow some occupation, I should certainly recommend Sydney as a place where they would be likely to spend their days as happily and profitably as anyone can expect to do who is banished from his native land. It might be well to spend the hottest summer months in one of the high-lying towns in the neighbourhood, or better still in Tasmania or New Zealand, but during the rest of the year I believe that Sydney could not easily be surpassed for a beautiful and healthy climate. The mean temperature of the summer in Sydney is 74° Fahr., and of the winter 55° Fahr., making the mean annual temperature about 65° Fahr. The hot north winds during the summer are very trying; and no great invalid should venture out of doors during their prevalence: they are succeeded by a wind from the south, which causes a very sudden and great lowering of the temperature, and which has to be provided against quite as carefully as the hot wind from the north. But no country in the world that I know of is free from all climatic disadvantages throughout the year: and every person, whether well or ill, is bound to take all reasonable precautions against them. To the best of my belief Sydney is as free from these disadvantages as any town in the world, with the exception perhaps of Hobart Town.

Inasmuch as Sydney is second only to Melbourne in size and population, and in its business relations with the rest of the world, whilst far superior to it in beauty of position and general attractiveness, I think that most travellers would agree with the opinion of its inhabitants, that it is *the* capital of the Australian colonies.

CHAPTER V.

QUEENSLAND.

Brisbane. — Maryborough. — Sugar-Making and Refining.— Coolie Labour. — Horse-Racing. — Australian Horses. — The Darling Downs.—Sport.—Snakes.—Sheep-shearing.—Climate.

WE left Sydney on the 15th of July, 1873, for Brisbane, in one of the steamers of the A.S.N. Co. We had fine weather and a good passage, and arrived at Brisbane early in the morning of the 18th. The town of Brisbane is beautifully situated on the river Brisbane, about 40 miles from its outflow into the sea. The river makes a sharp curve at that point, and surrounds the town on two sides. It is a fine river, being navigable for small steamers as far as Ipswich, which is about 25 miles above Brisbane. The banks are thickly clothed with unfamiliar trees and shrubs, and the country on both sides of the river is dotted with fields of cotton, sugar-cane, and other semi-tropical vegetation. The town itself is not beautiful. It consists of two long, straight, and broad streets, parallel to one another, and intersected by numerous shorter streets which connect them. The houses fronting the main streets are low and far from picturesque, and the town has the usual unfinished look of most colonial towns. The Houses of Parliament are large and fine, out of all proportion to the present size of the place, but evidently built with a view to the future, when Queensland will doubtless be a very important and prosperous colony.

There is a fair Museum, and good botanical gardens filled with tropical and semi-tropical plants and shrubs. Government House is prettily situated on an eminence above the town, but it is not comparable in size or appearance to the Government Houses of the other colonies which we visited. Brisbane contained when we were there rather more than 12,000 inhabitants, and it is increasing each year in size and importance. There are two or three moderate-sized inns in the town, but none of them looked very inviting. We were fortunate in not being obliged to try them. We had a letter of introduction to a gentleman in the place, and he immediately put our names down at the club, whither we sent our luggage, and went to stay on the first day we arrived. This club cannot compare in its appointments with the clubs at Melbourne and Sydney, but it was quite sufficiently comfortable; the only drawback that we felt at first was that our bedrooms were across a good-sized open courtyard, into which all the rooms opened, and which we had to cross each night when we went to bed. As it was mid-winter, and the nights were always cold and frosty, it seemed not improbable that invalids might find it somewhat risky to go out of hot sitting-rooms, through the cold night air, into the somewhat chilly bedrooms; but we felt no ill effects from this, as the air is very dry and still.

The climate during the first fortnight we were in Brisbane was exceedingly beautiful; the air light and clear without being at all too hot, and the nights bright, clear, and frosty. When we returned for a few days in the middle of August, the weather was much hotter; and when we returned again at the end of September, it was far too hot to be pleasant or comfortable in the day time; and the mosquitoes began to be very busy at nights. The inhabitants, who live, for the most part, in pretty villas in the different suburbs of the town, say that the heat, even during the summer months, is not unbearable, by reason of the dryness of the atmosphere; but they admit that the mosquitoes are troublesome. The heat must be considerable, as pineapples and bananas flourish in the open air; and it is too hot for grapes. I should say that no invalid ought to go to Brisbane, except during the months of June, July, and August; and I think that most persons would find that a short residence there would be quite sufficient for them. There are some pretty walks and rides to be taken in the immediate neighbourhood of the town, but the climate is not favourable for much active exercise; and there is very little to occupy or amuse a visitor in the town.

We were fortunate enough to make the acquaintance at the club of a gentleman who is very extensively employed in making refined sugar near to Maryborough, and he kindly invited us to pay him a visit. We accordingly accompanied him on his return home. We left Brisbane on the 31st of July in one of the steamers of the A. S. N. Co., and arrived at Maryborough early on the morning of the 2nd of August. Maryborough is about 250 miles north of Brisbane, and is situated some distance up the river Mary, the navigation of which is rendered extremely difficult by reason of extensive shallows and sandbanks nearly all the way up the river. At Maryborough we got into a small steam launch, in which we performed the rest of our journey. I may mention the name of our host, Mr. Robert Tooth, as his sugar-factories have a world-wide reputation for the size and completeness of their arrangements, and for the excellence of the sugar which he produces. The principal factory is situate about twelve miles up the river from Maryborough, at a place called Yengarie. Mr. Tooth's house was within a few hundred yards of the factory, and was surrounded by fields of sugar-cane, which he had brought under cultivation. The field work and all the unskilled factory work was done by natives of certain of the Pacific Islands, who were brought over under very stringent laws of the colony, which provide for the length of service of the men, their pay and food whilst employed, and their conveyance home at the expiration of their service. They seemed a perfectly happy and contented lot of fellows, with a friendly nod to us and a broad grin on their faces whenever they met us. Many of them were cannibals at home, and their wide open mouths and large, sharp, white teeth seemed eminently fitted for the purpose. We felt at first some slight apprehension lest a hungry fit should seize any of them on some night when we were asleep, especially as there were no bolts or bars to the outer doors, nor to any of the rooms of the house. The natives, however, seemed too well fed and cared-for to trouble themselves about human food; and we very soon got to be on excellent terms with them.

Much has been written and said against the so-called slavery of the Pacific Islanders in Queensland, and there may have been individual cases of hardship and oppression. So far, however, as our small experience went of the labourers at Yengaric and on the adjoining sugar-plantations, a jollier, more contented, and better cared-for lot of fellows I never saw. There is, however, one drawback to the system of employing Islanders in Queensland which I never heard mentioned by their friends, but for which the planters can hardly be held responsible. When they return to their homes after their terms of service they are not unfrequently killed, and sometimes eaten, by orders of their chiefs, who thereupon take possession of the valuables which they have taken back with them, and which they have purchased with their wages. These valuables are nearly all of the same kind; an old gun, probably one of our old military muskets, which are to be bought in the colonies for about twelve shillings each, and very dear at the price; knives, saws, and hatchets, with some little finery in the way of coloured handkerchiefs and beads.

A gentleman, who had been the Under Colonial Secretary in Queensland, told me that he went round the shops in Brisbane with the first batch of returning Islanders, in order to see that they were not cheated in their bargains. They had about ten or twelve pounds a-piece to lay out, and when they had bought all the things of the above kind which they wanted, they had each about fifteen shillings in hand, which they did not know how to spend. At last a bright idea struck one of them to buy a new green silk umbrella; and they all followed suit, and marched about the streets for the rest of the day, and started in the steamer next morning, under their umbrellas, with much apparent enjoyment. As they had no great amount of covering, except the umbrellas, the effect was most ludicrous to the bystanders.

Some of the labourers at Yengarie were pointed out to me who had been sent home after their first term of service, and who had afterwards returned to Queensland and re-engaged themselves for a second term; which does not look as if they had found much to complain of in their treatment. We spent nearly a fortnight at Mr. Tooth's house, and had a most enjoyable visit. We had fine, warm weather, not too hot, during the greater part of the time, and we daily had charming drives, rides and walks through the fields of sugar-cane, and the beautifully wooded bush country, which was gradually being brought under cultivation. We were shown over the factory, and duly initiated into the mysteries of making refined sugar out of the cane which we saw being cut in the fields: but these mysteries I am not going to divulge. The tropical vegetation, and the work which was being carried on all around us, were most interesting, and differed more from the experiences which we have at home, than anything we met with in the more temperate climates of the other Colonies. I must also confess that the mosquitoes were, in numbers, size, and ferocity, superior to any that we met with elsewhere, though the time of year when we were there was only just the commencement of the mosquito season. On our return journey to Brisbane we spent the day at Maryborough. The races were being held there, and the town was crowded with people who had come in from the country to see them. The hotels and lodging-houses were filled to overflowing with a most uninviting-looking set of so-called sporting-men; and it seemed to be altogether the sort of place where the shorter the time any respectable person had to remain the better he would be pleased.

I went to see the races, which were very moderate; but were made somewhat interesting to strangers by the way that most of the horses bucked and kicked, without in the slightest degree disturbing the equanimity of their jockeys, several of whom were aboriginal Australians. If there be one thing in which the Australians, whether white or black, or male or female, excel, it is in their horsemanship. Not that many of them could hold their own with our crack riders in a hard run over a difficult country in England, as they are not accustomed to that kind of riding: but for sitting calmly and easily some of the greatest brutes of horses, while they go through every possible manœuvre to get rid of their riders, I never saw the equal of the aboriginal Australians, and next to them, of some of the young Australian squatters, who almost live on horseback, and who all break their own horses. These young men used to look down with great contempt on us when we begged for the quietest horses on the station; and many of them seemed really to prefer a buck-jumping horse to one that went along soberly. As some of these horses can buck out of their saddles without breaking the girths, it will be seen that it requires some skill in the riders not to be left behind, or shot over their horses' heads, during the operation. It is considered by some of the least amiable of these young men a great joke to put a "new chum," as they are pleased to call new arrivals in the country, on one of the well known buck-jumping horses on the station; and they think lightly of such small matters as a broken arm or collar-bone as the result. Thanks, however, to our numerous hosts in the different colonies, who were all of a very different stamp from those to whom I have referred, we were always given quiet, steady-going horses, which required no horsemanship, and which were remarkably clever in scrambling in and out of the many curious places which are met with in every day's ride. It is quite a mistake to suppose that all Australian horses are buck-jumpers. There are many that can hardly be beaten for steadiness and staying powers during long journeys of hundreds of miles in the bush, and which seem to live almost entirely on water and a roll in the sand when the saddles are taken off at night. They are allowed to drink as much water as they like at every creek or water-hole during the journey; and their paces are walking and cantering, which they will keep up for days and days together, doing on an average forty to fifty miles a day without a bit of corn.

On the country stations no one ever thinks of walking if he can possibly help it. I have seen men work hard for an hour in the hot sun "running in" a horse in order to ride half a mile; and it is almost the only means of progression for long distances. So averse indeed are the young squatters to walking, that when they are staying in the large towns on business or pleasure, they will jump into a hansom to go half way down the street. This habit of riding everywhere on the country stations is a great advantage to invalids, as horses are provided for every one as a matter of course; and horse exercise can be indulged in every day and nearly all day long, with very little fatigue or exhaustion.

We did not go any further north than Maryborough; nor do I think that, from a health-giving point of view, any good would be likely to be gained by doing so.

The next important town of Queensland north of

Maryborough is Rockhampton, which is in the tropics; and such is the climate of Rockhampton, that it is a common saying in the Colonies that if a man had two houses, one there and one in the infernal regions, he would let the former and live for choice in the latter. We left Maryborough early on the morning of the 15th of August on our return to Brisbane, and we arrived there on the following morning. We only remained there three days, as we had letters of introduction to two or three gentlemen who had large sheep stations on the Darling Downs, and who had kindly invited us to pay them a visit.

The Darling Downs is the name given to a large extent of table-land about 2000 feet above the sea-level to the north-west of Brisbane. There is a good coach-road from Brisbane to Ipswich, which is at the foot of the Downs; and the rest of the journey is made by railway. The railway journey is very fine. The line winds gently up the steep incline from Ipswich, amongst densely wooded hills till it reaches Toowoomba, which is one of the principal towns on the Downs. The line then continues on the level through the broad plains of the Downs as far as Warwick, which is the other principal town; and there are railway stations or sidings for all the principal sheep stations along the line.

The station to which we were going was about fifteen miles beyond Toowoomba, and we arrived there just at sunset on a beautiful clear evening, cool enough to enjoy a fire when we arrived at the house. It would be no advantage in a book of this kind to give the names of the stations which we visited, nor of the owners who so kindly entertained us; and I think it better to avoid mentioning them. It is sufficient to say that we spent a month most pleasantly amongst three large sheep stations between Toowoomba and Warwick, and that we were most kindly and hospitably entertained at all of them by the owners, who were gentlemen who had never seen us before.

As the life was similar at each of the stations, I will give a slight sketch of it, which is applicable to all of them. All our hosts were married men with families: and they had comfortable houses, well furnished, and with all the conveniences of civilised life.

We heard a great deal at different times about "roughing it in the bush;" and no doubt there is plenty of roughing to put up with on many stations through the different Colonies, especially where there are no ladies to keep the owners up to the proper standard of civilised life; but I had no personal experience of it. Nearly every house at which we stayed had well furnished dining and drawing rooms, with three or four comfortable bedrooms, and a large verandah in which a great part of each day was spent, and in which beds were laid down for unexpected guests at night when the house was full.

The great charm of the life is its perfect freedom. No one interferes with what the guests like to do, so long as they conform to the ordinary hours of the household. There are always horses to ride or drive; or the visitor may roam about with his gun for miles without the slightest interruption. In the latter case it is well for him to be provided with a compass, and to take the bearings of the house before he sets out. If it had not been for this precaution I should have spent a great many hours, and possibly the whole night, in the bush at the very first station which we visited. I had walked down a winding stream, or creek as it is there called, to look for ducks; and after walking for three or four miles without much sport, I thought I would return by a short cut away from the creek. The fine morning had changed into a thick drizzling afternoon, but I had no doubt about the direction of the house. I walked for nearly two hours on what I felt sure was the homeward track, but I did not come within sight of the house; and eventually I pulled out my compass, which showed the position of the house to be nearly at a right angle from the course which I was taking. I fortunately trusted to the compass, though I could hardly believe that I had taken my original bearings correctly, and followed the course which it pointed out, and which eventually brought me home before darkness set in. I was two or three hours late for dinner, and found my host very much disturbed at my absence and on the point of sending out men on horseback to look after me. The features of the country are so much alike, that it is difficult for a stranger to identify the landmarks with which those who are acquainted with the district become so readily familiar.

There is plenty of sport to be got on the Downs at certain times of the year. Large herds of kangaroos are to be found everywhere by the hunter, and smaller game is fairly plentiful for the shooter. Flocks of wild turkeys are met with in some districts, and wild ducks of many different kinds abound in the creeks and water-holes. Snipe are found in some parts, and plenty of quail in the right season. There are also many kinds of pigeons of beautiful plumage, and exceedingly good to eat, in all the wooded parts of the Downs, and large flocks of parrots and white cockatoos, which are also considered great delicacies if properly cooked. There are, on the other hand, snakes of different kinds and in great quantities, the bite of all of them being more or less deadly, the palm in this respect being given to an ugly beast called a deaf adder, which either hears not or heeds not the approach of the unwary traveller. We were told that within half-an-hour of being bitten by one of these adders, the person bitten is past all hope of recovery by any known remedial measure; and, indeed, that no remedy has hitherto been discovered of the slightest use, even if applied immediately. An injection of ammonia into the veins has been found efficacious in the case of bites of most of the Australian snakes; or failing this, a bottle of brandy or whiskey taken internally. If the bitten person can be made intoxicated before his blood is thoroughly poisoned, he is safe; but this is a difficult thing to do after a snake-bite, and it is of no use whatever in the case of the deaf adder. Fortunately this adder is not common, and nearly all the other kinds of snakes glide away at the least sound, and there is thus not much fear of treading on them.

It is strange how few bad accidents occur from snakes in a country which swarms with them to such an extent as does Australia. In the preceding winter to that in which we visited the Darling Downs, there were heavy floods, which did great damage amongst the flocks of sheep. A flock of about 1,100 were drowned on one of the stations which we visited; these sheep were skinned, and the skins left on the ground to dry; when they were removed some time afterwards, a snake was found coiled up under nearly every skin, having gone there for the sake of warmth, and these were of course all killed. During the same flood over 400 snakes were killed on the adjoining station, having got up into the fences to escape out of the water. The owners declared that after this wholesale slaughter there was no sensible diminution of snakes on their stations, and there was certainly a good supply on both stations when we visited them.

The Darling Downs are famous for their sheep stations, and we were initiated into the mysteries of washing and shearing sheep on the most improved principles and on a very large scale. Many of the large Australian sheep-owners, squatters as they are called in the Colonies, shear from 50,000 to 60,000 sheep in each year, and some of them more than double that number. I am not a judge of wool, but I brought some specimens home with me, which I took at random from some of the sheep on the Downs which were standing ready to be shorn, and I am told on competent authority that they are of exceptionally fine quality.

Different methods of sheep washing and shearing are employed by the different squatters. Some, who have a difficulty in procuring sufficient water power, prefer to shear their sheep in their dirty, or **greasy**, state, and then forward the wool to be cleaned before sending to market. Where, however, it can be done, the other method, namely that of washing the sheep before shearing, is preferred, and the wool commands a higher price in the market.

It was a curious and most interesting experience to ride off after breakfast with our host to the wool-shed, which is always placed in the most convenient situation on the station, and often five or six miles from the house. We saw the sheep being driven up toward the shed in flocks of 500 or 600 at a time. These were packed close together into an enclosure, surrounded at a little height with iron pipes, which were perforated all over with very small holes. The water was forced into the pipes, and came down through the holes in the form of small rain on the sheep for several hours together. The sheep were then taken and washed with soft soap and warm water, and afterwards shot down an inclined plane into a sheep pool, where they were pulled about, ducked and rubbed by men who were standing in tubs in the pool, until they were finally pushed out half drowned at the other end of the pool. It cannot be a pleasant occupation for either the washers or the washed, but it went on from sunrise to sunset each day for many weeks together. The sheep were then turned into a clean paddock, from which they were driven into the shearing-shed, whence they emerged in often a sorry-looking condition, being naturally rather thin, and being generally more or less gashed in the operation of shearing. Shearers go round in gangs to the different stations at the proper time of the year, and the skill of some of them is astonishing. I was told that many of them could shear a sheep thoroughly in five minutes after they got their hands accustomed to the work. It took an average of seven or eight minutes when I was on the Downs, but it was then in the first days of the shearing, before the men had got their hands in. Many can shear with both hands; they will go down one side of the sheep with the right hand, and up the other side with the left hand, and thus save a great deal of time and trouble. They are paid by the score or the hundred, so that their great object is to shear as many as possible in the shortest time, and this naturally results in gashes and scars on the backs of the unfortunate sheep. As, however, each shearer's lot of sheep is turned into a separate pen, the overlooker can see at a glance who is the offending gasher, and a deduction is made from his pay for each gashed sheep. It is probable, therefore, that, in the long run, the shearers find it pay better to shear rather a smaller number of sheep carefully than a greater number slovenly; and the later-shorn sheep come in

F

for better treatment than the earlier ones under the practised hands of the shearers. I have seen above thirty shearers employed in one shed; and I believe that on one or two stations there are sometimes between eighty and one hundred men employed at the same time. There was one squatter mentioned to me as shearing nearly one million of sheep on his different stations throughout the colonies.

I have referred, perhaps at greater length than may be thought necessary for the purposes of this book, to some of the manners and customs of the residents in the colonies, but I venture to think that these details are not out of place. One of my objects is to explain the different interests and amusements which the invalid traveller may meet with on his travels; and no one can stay long at the house of an Australian squatter without taking a great interest in the subjects which are of such vital importance to his host. It is a real pleasure to me to dwell upon the recollection of our month's visit on the Darling Downs. We had four wet or partially wet days; and on two other days there was a high and cold west wind: but every other day was magnificent. A bright sun, with only sufficient cloud to contrast with the deep sapphire blue of the skies, and with generally a fresh air to temper the rays, shone down upon us day after day. It was neither too hot to ride about the stations in the morning, nor too cold to smoke our cigars in the verandah in the afternoon; whilst the nights were clear, bright and frosty.

The climate at that time of the year was, in my opinion, the finest we experienced in Australia; and I certainly derived great benefit from the open-air life which we lived. We were told by some people that it was unbearably hot on the Downs in summer, and very cold and sometimes wet and windy in the height of winter: but the residents assured us that there were no great extremes of either heat or cold. I could fancy that in a wet season the earth might retain the moisture for a long time, as it is a deep black soil, rich enough to grow anything, though the Downs are at present almost exclusively occupied as pastoral land. I should think that with ordinary care the Darling Downs might be made a pleasant and healthy residence during a great part of the year. There are fair inns at Toowoomba and Warwick, though the charges at the latter seemed somewhat excessive, considering the accommodation. I should prefer Warwick to Toowoomba, as being built in a more healthy situation; but life in either place would be intolerably dull. I could not recommend any invalid to go to the Darling Downs with the intention of making a long stay there, without one or more letters of introduction to the owners or occupiers of the stations on them.

We returned to Ipswich on the 24th of September, and started early in the morning of the following day by steamer for Brisbane down the river Brisbane, the scenery of which is exceedingly pretty. We remained in Brisbane for a few days only, and started on the 1st of October on our return journey to Sydney.

I have a great idea that, in the course of a few years Queensland will be one of the richest and most prosperous of the Australian colonies. There is now a special service of mail steamers between London and Brisbane, the Eastern and Australian Steam Navigation Company, which, passing through Torres Straits, brings Queensland into direct communication with Europe and Asia, without having to go round three sides of the continent of Australia. It is a shorter and very interesting journey from Galle to Brisbane by this line, though by no means wholly free from difficulty or danger by reason of the intricate nature of the navigation through the Straits. The position of Queensland, being partly in and partly out of the tropics, and thus combining the advantages of tropical and more temperate climes and containing the vegetation of both, is admirably adapted for the successful carrying on of numerous industries, and notably those of sugar and cotton, in addition to the great Australian sources of wealth, the sheep and cattle stations, which are common to the other colonies.

We arrived in Sydney on the 3rd of October, and remained there until the 4th of November, when we went forward to Melbourne, arriving there on the 6th. I remained in Melbourne till the 10th, when I went on a visit to a friend who had a large sheep station in the Riverina. As I greatly enjoyed, and very much benefited in health by, my long visit at this station, I shall add a chapter giving my experiences of the Riverina.

CHAPTER VI.

THE RIVERINA.

Name and Situation.—Journey from Melbourne.—Bush Scenery.—Station Life.—Salt Bush.—"Nipping and Shouting."—Bush Inns.—Wild Turkeys and other Game.—Kangaroos.—Scarcity of Water.—Horse-breeding.—Travelling in the Bush.—Wagga Wagga.—Races.—Visiting.—Keeping Christmas.—Great Heat.—Climate.—Homeward Bound.—Australia for Invalids.

THE large district called the Riverina derives its name from being situated amongst the four large rivers Murray, Murrumbidgee, Darling, and Lachlan. It is bounded on the south by the Murray; but for all its other boundaries, and for an excellent description of the country, its inhabitants and products, I must refer my readers to Mr. Anthony Trollope's interesting book called "Australia and New Zealand," which seems to me very accurate in description, and most useful in giving a fair idea of the respective colonies and their capabilities. The Riverina comprises a very large extent of pastoral country which forms part of the colony of New South Wales; but it is situated at so great a distance from the capital, Sydney, that the inhabitants find it cheaper and more convenient to transmit their wool and other produce to Melbourne, rather than to their own capital. Consequently the squatters, who form the aristocracy of the Riverina, as they do of the rest of Australia, are more frequently to be met with on business or pleasure at Melbourne than in Sydney. I had the good fortune to make the acquaintance of a Riverina squatter in Hobart Town during the preceding summer, and he kindly invited me to pay him a visit. He was at Melbourne for the race-week when we returned from Sydney, and I arranged to accompany him on his return to his station. I shall not mention his name here, but he is well known for his great kindness and hospitality to all English visitors in the colony, and his identity will be easily recognisable by any person in his district who may happen to see this book. I shall only add here one word of hearty thanks to him, and to all our other kind friends in the different colonies, for making our visit to Australia so agreeable to us, and so profitable from a health-giving point of view.

The journey from Melbourne to the Riverina was at that time made by rail as far as Echuca on the river Murray, and thence by coach to the different townships of the district. It was in contemplation to continue this railway as far as Sydney, but there were at that time intercolonial jealousies between Victoria and New South Wales which prevented it being carried out. These jealousies have now been set at rest so far as to permit of this most important work being completed, and the line connecting Melbourne and Sydney is now made and opened for traffic. This line passes through a great part of the Riverina, and it will make the country much more accessible from either of these capital towns.

We left Melbourne on the 10th of November

by an early train, and dined at Echuca, and started on the coach for Deniliquin in the afternoon, arriving there about eight o'clock in the evening. It is not a beautiful or interesting drive: the road runs partly through interminable grass plains, which at that time of the year were quite brown, and partly through low scrub and bush country, the trees being just high enough to prevent anything being seen beyond them. We spent the night at a very comfortable inn at Deniliquin, where the coach stopped; and we started by coach at six o'clock next morning for Jerilderie, arriving there about three in the afternoon. It was a pleasant drive of about sixty miles though open plain country, but it cannot be called beautiful. We spent the night at Jerilderie, which is about forty miles from my friend's station, and on the next day we went forward in his buggy.

The country was similar to that which we had passed through on the preceding day. Enormous grass plains, the colour, at that time of year, of English stubble-fields, enlivened occasionally by plots of greener grass and clumps of good-sized trees, wherever the creeks had either naturally or artificially irrigated the surrounding country. Wherever there was water we saw wild fowl of different kinds; and as we got nearer to the station we saw large flocks of wild turkeys, giving promise of sport in the future. My friend had a nice house, and a pretty garden with plenty of flowers and vegetables still flourishing in it, though the garden was completely burnt up and changed into a desert long before the end of my visit. He was a married man, and his wife had two nieces staying with her; so that, with a friend who had come with us and the manager of the station, we had a large party in the house. Moreover, it often happened during my visit that some young neighbouring squatter turned up unexpectedly to revive his memories of civilised life, after spending weeks or months by himself in the Bush.

Life on a station in Australia must be rather monotonous, even under the most favourable circumstances, and would probably be called dull by people fond of society and with no resources of their own; but to most men who are fond of outdoor pursuits, especially if not in very robust health, the life is full of attractions. The house is a veritable "liberty-hall." Each person does what he likes so long as he conforms to the ordinary domestic arrangements of the establishment. I was out in the open air all day long. We walked or rode about the place all the morning; we rode or drove in the buggy after lunch, and very often with a gun in the buggy, with which we could always get as many wild turkeys as we wished; we played croquet from about 5 P.M. till 7, when we dined, after which we sat in the verandah with our cigars, and finally wound up with a rubber of whist in the drawing-room, with doors and windows wide open till bedtime. This, with pleasant society and excellent food and liquors, suited my case entirely, as I think it would that of most people in a similar state of health. The post only came in and went out once a week, when one of the men-servants had to ride about forty miles to the nearest office; and at that time there was only a mail from and to England once a month; so that, except on the day preceding the post day, there was no writing of letters to be done, and telegrams were almost unknown, neither was there any daily paper to be got through.

It is wonderful how, after a short residence in the colonies, European politics cease to have any vivid interest even to those who are most interested in them at home; and we were quite satisfied with the short *résumé* of news contained in the *European Mail* or *Home News* which arrived once a month with the English letters. There were no callers in the English sense of the word, nor any tradesmen coming round for orders; but all the stores had to be brought from Melbourne once a year in the return bullock-waggons which had taken down the wool for sale in the spring. The nearest neighbours were from thirty to forty miles distant, and it required a long experience of the Bush to hit off the right line to their houses. My friend's station was a sheep station, and the principal article of food was mutton, and excellent mutton; but this was varied by poultry and any game that was in season. When I was in the Riverina there was hardly any game to be got except turkeys, but all kinds of wild fowl abound in the winter, and make a most excellent addition to the larder. The country is what is there called salt-bush country, so named from a low shrub with a very strong salt flavour in the leaves, which grows in patches throughout the district, and is most valuable as food for the sheep and cattle, especially when the grass becomes dried up and wanting in nutrition. The milk of cows which eat this salt-bush is considered particularly beneficial to persons suffering from lung complaints, and a large jug of this delicious fresh milk was, by the instructions of my hostess, always standing in a cool place for me to drink at any hour of the day, qualified sometimes, by the advice of my host, with the smallest possible nip of whiskey. This is a drink much to be commended for its palatable and nourishing qualities. The ordinary drink of squatters at meal-times is water, or tea where the water is not good, and they take their alcohol in the form of whiskey or brandy between times. This is a bad system, and one apt to be abused, as any time of the day after getting up is considered suitable for a "nip," and what is there called " a full-grown man's nip," means a good-sized wineglassful of spirits, with about twice as much water. This, again, sometimes leads to unpleasantness between those who are fond of nips and those who are not; as it is considered the height of rudeness to refuse to be treated to a glass of spirits when asked, which of course entails a return of the compliment at the first convenient time. It is particularly troublesome on a coach journey, where it is thought necessary, in the interests of the Bush innkeepers, to drink at every little inn where the coach stops; and each passenger treats, or as it is there called "shouts for" his fellow-passengers in turns. I found it simply impossible to fall in with this custom, and I was sometimes looked upon as rather a poor fellow accordingly.

The result of this bad custom is already beginning to tell seriously against the colonists, and especially the younger portion of them. There is a deal too much drinking throughout the colonies, and more especially in the Bush, away from the towns. There is no doubt a great temptation to those who are fond of drink to indulge in it when living a solitary life in a small wooden house in the Bush, without any of the ordinary subjects of interest or amusement, and especially after a hard day's work in the open air. These solitary individuals have probably read through every scrap of printed matter which they may ever have had, and they find it hard to kill the time between sunset and bedtime. Not unfrequently, too, a friend similarly situated on a neighbouring station, will ride in unexpectedly, and the only delicacy which his host can offer him is "Hennessy's Three Star Brandy," or "Kinahan's fine old Irish Whiskey." Then they make a night of it. Whatever may be the cause, however, it is a fact that the habit of drinking is a very serious curse in the colonies, and any invalid who goes there must set his face steadily against it from the very first, if he intends to get any benefit from his residence there. I need hardly say that I never saw anything of this kind on any station at which I stayed, as my hosts were all educated gentlemen, with wives and families, and the liquor provided was just what we get at home;

sherry wine, and the wines of the colonies, some of which are excellent, with good claret and English beer and porter, were given us, even in houses where they were not ordinarily drunk, in consideration of our being English guests, and my being an invalid; and as every bottle had to be brought up from the nearest large town, often 300 or 400 miles distant, some idea can be formed of the hospitality of our hosts.

The bad habit of drinking too much naturally descends from the richer to the poorer classes, and by reason of it, many a man who works all his life on a station, with good pay and no expenditure for food or lodging, is just as poor at the end of his life as at the beginning. No person employed on a station is allowed to bring spirituous liquors to his hut, and any breach of this rule is usually followed by instant dismissal. The men contentedly drink water or tea during fifty weeks out of the fifty-two, but at some period of the year, commonly about Christmas, they ask for a fortnight's leave of absence and a cheque for their wages for the whole year, and go straight to the nearest Bush inn, where they give their cheque to the landlord with instructions to supply them, and any friends who may happen to turn up, with food and liquor until the money is spent, or, as it is there called, until the "cheque is knocked down." The landlord usually keeps two kinds of liquor, one for the gentlemen and one for the working-men, the latter a vile compound of, apparently, the strongest spirits of wine and molasses; a few glasses of this are sufficient to make the customer perfectly drunk and unconscious, and in this state he will be kept for perhaps ten days or a fortnight, when the landlord kicks him out, telling him that the cheque is knocked down. Cases are not unknown where the customer, after being kicked out of the inn, lies down in the bush in a semi-unconscious state and is found dead in a day or two. But the usual course is that, after two or three days of sobering, the man returns to the station, and works steadily and well on water or tea till his drunken fit comes round again at probably about the same period of the year. And this goes on year after year during his whole lifetime. There are some squatters, with a not very high standard of morality, who rather encourage this miserable habit, as it keeps their men, who may in other respects be very competent workmen, in their service, and prevents their saving money and rising to be small farmers, or free selectors, on possibly their own station; but the better class of squatters do their best to break their men from this habit, and urge them to save their wages, or at any rate some portion of them, each year. It is, however, very seldom that they succeed. Partly their own desire for a spree, and partly their wish not to appear mean in the eyes of their friends and of the innkeepers, who will hardly take them in unless they bring a cheque for the whole year's wages, cause the men to stick to their old habits. And it is very seldom that one sees a workman on a station who has saved any considerable part of his wages, no matter how respectable the man may look, nor how long he may have been working satisfactorily on the same station.

To those who are fond of shooting, there is plenty of sport to be had in the Riverina with wild fowl and snipe and quail at the proper time of the year. I was, unfortunately, there out of the season for nearly everything but turkeys, but we had great fun driving after them. They are so shy that no person on foot can approach within a quarter of a mile of them, but they will sometimes allow a person on horseback to get pretty near them. The best way, however, of shooting them is for two persons to drive up to them in a buggy, one holding the reins and the other the gun. If this be well managed, the sportsman can generally get within easy distance of them. They are a species of bustard, and very good to eat. The flesh is red, with a decided flavour of game; but they are much better in winter, when they can be kept for some time after being killed. In the summer it is often so hot that game is unfit for food by the time that the sportsman gets it home after shooting it. There is another kind of turkey, called the brush turkey, from living in bush country, as this one is called the plain turkey, from living on the plains. The flesh of the brush turkey is white, and not unlike that of our domestic turkey, and it is almost as good for eating purposes. The average weight of both kinds of turkey is about 12 lbs. in their feathers, but they sometimes reach to 16 lbs. and 17 lbs., and sometimes even to 20 lbs. in weight.

There are large flocks of kangaroos to be seen all over the country, and if your host should happen to keep kangaroo hounds, you may have many a good gallop with them over the plains. The paddocks in the enclosed country are seldom less in extent than 10,000 acres each, so that it is possible to have an uncommonly quick run and kill without a check. If, however, the kangaroo can get to a fence, he is generally safe, as I have seen them fly over a six-foot post and rails with the greatest ease, and very few horses can follow them. Then there were the sheep in distant paddocks to be looked at occasionally, so that we were never in want of an excuse for riding about and living in the open air.

One of the curious things that every new-comer notices, is the apparent absence of sheep on these large sheep stations. The paddocks are so large, and the whole country on so vast a scale, that it is almost impossible to believe that the few sheep which one sees dotted about at great intervals, can make up the 50,000 or 60,000 which is about the average number on these Riverina stations.

The country about my friend's station could not be called picturesque. There were miles and miles of level plains, hardly broken by a single hillock, and with but little timber upon them. There were occasional clumps of trees in places where the water stood in the winter-time, but the general aspect during the summer was flat and bare, and of the colour of an English stubble-field. The plains were green enough in the early spring, when the grass grew to the height of three or four feet; but when I saw them, I wondered how the sheep and other animals could possibly have found nourishment enough for subsistence. There was no river, or even large creek, running through my friend's land, but the sheep and cattle were watered by means of deep wells, which were sunk on different parts of the property, and the water from which was drawn up by means of whyms worked by horses, and pumped into long troughs night and morning. This entails great expense at first, but it has the advantage that, in seasons of great drought, when the creeks are dried up, a fair supply of water can nearly always be obtained from the deep wells.

My friend was a real sportsman, and was one of the principal amateur breeders of thoroughbred horses in the colony. He had then achieved many successes with his racehorses; and since that time he has become still more successful, and has won some of the principal races in the different colonies. It was a beautiful sight to see the brood mares with their foals coming up to water every evening a little before sundown; and we used to amuse ourselves by picking out the winners of races in the future.

Soon after my arrival in the Riverina, the race-week was held at Wagga Wagga, which is about 100 miles from my friend's station; and as he was running some of his horses there, it was arranged that our whole party should drive over there for the week. I will give a short account of our expedition, as it is typical of many similar expeditions in the Colonies. Many of the large squatters keep a pair of horses for their own buggy, and perhaps a riding-horse in a stable near at hand, and these are better cared for and attended to than the rest of the station horses, which are never groomed, and which have to find their own living. None of the horses in the country districts are shod; and one seldom sees any lameness arising from injury to the feet. When an expedition has been determined on, the first thing to do is to "run in" a lot of horses, and pick out those which may be considered suitable for the purpose. A team of three or four horses is often put together, no one of which has ever been in harness with any of the others, and may possibly have never been in harness before. It is then that the skill which all Australian squatters have with horses is seen: the team may go as awkwardly as possible at first, but go they will, and generally fairly well, after a couple of hours of uncomfortable travelling, not always without damage to the vehicle or harness. A hatchet and coil of rope are always taken in case of accidents, and they are generally wanted; but it is not often that there is much injury to the lives or limbs of the travellers.

We were a party of six, and we started in two vehicles; one, the buggy with two horses, in which were the three ladies driven by my host, and the other an American waggon with three horses, carrying the luggage, with a high driving seat on which I and a friend of my host sat, he driving a team which we had " run in " on the preceding day. There was no visible road, but we drove quietly on across the plains without any adventures till we came to a broad stream, called the Yanko Creek, which had to be crossed. It looked dark and deep, and one of the servants who was riding with us was ordered to try the depth at the ordinary crossing. His horse was soon off its legs, and the man swimming behind it, holding on fast by its tail, which is there considered one of the safest modes of crossing a river on, or rather off, horseback. Man and horse got safely to the other side, but we saw that the water was too deep for the carriages, so we drove back to a neighbouring station, which was on the creek, and the owner of which had a small boat, where we crossed in a very original manner. The horses were taken out of the vehicles, and the cushions out of the buggy, and the luggage out of the waggon. Ropes were then made fast to the buggy, and one end carried by men in the boat across the creek and passed round a tree on the other bank, and the buggy was dragged forcibly through the stream, two empty barrels having been first fastened underneath to make it buoyant. The buggy disappeared entirely from view under water; but fortunately there were no snags or sunken trees at the bottom, and it came up safely on the other side. The same process was applied to the waggon with equal success; and we and the luggage then crossed over in the boat, the horses swimming alongside. The passengers, cushions, and luggage were then stowed back in their respective places, and we drove on as if nothing out of the ordinary course had happened. We were threatened with a similar difficulty at another creek on our way, but fortunately the water went down during the night, and we drove through the ford next day without accident. We spent the night at a place called Gillenbah, and arrived at Wagga Wagga about six o'clock on the next evening. The scenery could nowhere be called fine, but it was rather pretty during some portion of the way along the river Murrumbidgee.

Wagga Wagga is a small country town of about 2000 inhabitants. It will be remembered in England as the residence for some years of the unfortunate baronet who is now wasting his years in prison. He followed the respectable trade of a butcher in the town, and his shop was pointed out to me, with the name "Thomas Castro" still legibly painted on the door. The subject was of greater interest when I was there than it is now, and I made inquiries amongst his former neighbours as to what they thought of his claims. I could find no one who had the slightest belief in the validity of them; and they all seemed astonished that he had so large a number of professing believers in England. The shop was then about to be pulled down to make way for newer buildings, and the claimant and his claims are no doubt as much forgotten by this time in Wagga Wagga as they are in England. It is rather a pretty town, and the neighbouring country is not so flat as the country through which we had driven. It is one of the principal towns of the Riverina, and there are several good-sized inns or hotels in the place. We were at the Australian Hotel, which was comfortable; but the town and all the hotels were crowded with visitors of all kinds, including a large sprinkling of betting-men very similar in appearance to those we have at home. The race meeting does not merit any particular description, but we had three days of moderate racing, and I made the acquaintance of all the rank and fashion, and youth and beauty of the district.

Horse-racing is *the* national sport in Australia. Every town, with above a certain limited number of inhabitants, has its race-course and Grand Stand, and generally contrives to hold one or two meetings during the year. I fear that the morality of the turf is no better in the Colonies than at home; and I saw many horses which ran like rogues in small races, and like trumps in more important ones, when the money had been piled pretty well upon them. Fortunately, however, there are gentlemen in both countries who breed horses in the true interests of sport, and who run them in a similar spirit; and these are naturally more numerous in the Colonies, where the opportunities for rearing and training horses by amateurs are so much greater than at home. Every one in the country districts of Australia is brought up on horseback from early childhood, and whenever three or four young men meet together on some station for two or three days' "spell," or freedom from work, they are sure to get up races amongst themselves, and are prepared to back their horses, or their own skill in horsemanship, for any reasonable amount of money. This has its bad, as well as its good side, as many of the young men get a horsey, not to say slangy, manner about them, whilst their ideas and conversation are almost entirely confined to the subject of horses, and their inventive powers are seldom at a loss in describing their individual, and sometimes wonderful, feats of horsemanship.

We returned to my friend's station from Wagga Wagga without any adventure, the creeks having subsided, and admitting of being crossed at the fords without much trouble or difficulty.

The weather in the Riverina during the whole month of November, and up to the end of the first week in December, was beautiful; but after that time it became too hot for comfort, and even for health. It would probably have been wiser if, on the latter grounds, I had left the station then and gone straight to Tasmania; but my friends were so urgent in their hospitality that I should remain with them until they went there, which they proposed to do after Christmas, that I consulted my own inclinations and stayed on.

One day, early in December, my friend and I started in his buggy to make a tour of two or three of the neighbouring stations. We had a hot, but not unpleasant, drive to the first one, about forty miles distant; but on the next day a hot wind sprang up, which quite prevented our going forward. The effect of a real hot wind at that time of year is horrible. The plains were hidden in a continuous cloud of dust, and the air felt as if it came direct from a furnace. The thermometer in a shady and protected part of the verandah at the station to which we had driven, stood at 105° Fahr. all the morning, and we spent the morning in a perfectly darkened room, with every door and window carefully closed, and we all felt in a complete state of exhaustion. In the afternoon a south wind came up strongly, with black clouds and a thunderstorm, which immediately lowered the temperature to about 80°, but the air never recovered its freshness, and it remained hot and sultry all the evening and night. The next day was fine and clear, but very hot, and we went forward to another station, a distance of between forty and fifty miles, and on the next day returned to my friend's station. This is the free and pleasant way in which squatters keep up social relations with one another. They drive up to their friend's door without any previous notice, and are gladly welcomed as a matter of course. The groom takes the horses to the stables, or if there be neither groom nor stables, which is a not unfrequent occurrence, then the squatter and his friend take off the harness, water the horses, and turn them out into the home paddock. They then go into the house, where spirits and water are produced, and, if smokers, a friendly pipe smoked, whilst the ordinary subjects of interest connected with the respective stations are discussed. By this time the bedrooms have been prepared, or if they happen to be already filled, the bath-room, which is attached to every moderate-sized station, has been made tidy, and a good wash clears off the stains of travel. The visitors can then make their appearance before the ladies, if there be any. No extra preparations are made in the way of entertainment. If it be a sheep station, there is sure to be plenty of mutton, and if a cattle station, plenty of beef; and the soul of the visitor is not vexed by the anxious face of his hostess, lest there should not be enough to eat, nor with the appearance at table of ill-made side-dishes, which is so often the case at home. If the visitor cannot have a bedroom, he has a comfortable mattress in a shady part of the verandah, with the bath-room, in which to perform his toilet. And this leads to real friendly unceremonious visiting. The too frequent phrase, "Not at home," is unknown in the bush, unless such be the real fact, and even in that case, the visitor thinks nothing of remaining all night, as his next stopping place may probably be forty or fifty miles further on.

On Christmas Eve four young squatters who had ridden in from neighbouring stations, turned up unexpectedly to spend their Christmas with my friend. These all had their quarters in the verandah. It was a very hot evening, but we made ourselves still hotter by having a snap-dragon, by way of keeping up the traditions of the old country. Christmas Day was again very hot, the thermometer standing at 101° Fahr. in a shady part of the verandah during a great part of the afternoon. We had, however, a regular Christmas plum-pudding for dinner, brought in blazing in the most approved fashion, and very hot it was. We had begged to have the haunch of mutton roasted earlier and allowed to get cold. It was a curious sight to see us all in white coats, and the lightest garments of other descriptions, trying to think we were keeping Christmas at home, and drinking toasts to all our absent friends. The word "home" is the simple and touching expression which all colonists still use when speaking of England—and long may they do so. The next day was rather cooler, and the younger guests got up flat and hurdle races on horseback, some of them also doing curious feats of fancy horsemanship, such as are done in a circus in England, but so much the more difficult in that they were done on the open plain, on half broken horses, and without the accompaniment of slow music. The guests went home on the next day, and then followed a succession of terribly hot days, varied occasionally with a hot wind, until the 14th of January, when we left the station. During this time the thermometer registered every day over 100° Fahr. in the shade at some time during the afternoon. Once it registered 114° without any hot wind, and on the following day 110°, and on several days 105° and 106°. The nights also were very hot, and the sheets, when we got into bed, felt as if they had just been warmed. We could do nothing but lie on the drawing-room floor and sleep during the greater part of each day. The windows and doors of all the rooms were closed carefully from morning to near sunset, so as to exclude the hot air; and every one on the station became more or less out of sorts and unable to enjoy life.

My brother remained in Melbourne when I went to the Riverina, but the heat was so great in the town during December that he could not stand it, and he went off to Tasmania before Christmas. We heard that after Christmas the heat and dust in Melbourne were almost unbearable; the ranges of the thermometer were nearly as high there as in the Riverina, and the excessive heat in the town was much harder to bear. We had fortunately rather cooler weather for our three days' journey to Melbourne, and it was not quite so hot in the town when we arrived as it had been. It was, however, quite hot enough, and we were glad to leave again the next day for Tasmania. We set sail on the 19th of January for Launceston, and drove thence to Hobart Town by coach across the island, as on the previous occasion.

The Riverina has a magnificent climate during nine months of the year, and one particularly adapted for persons suffering from affections of the lungs, but it is far too hot during the months of December, January, and February to be advisable as a residence for invalids. Those of the inhabitants who are able to do so, make a point of going away for a change during some part of that time, and it is then that Tasmania and New Zealand become favourite places of resort for persons seeking health and pleasure.

There are many other districts, both in Victoria and New South Wales, which would make excellent health resorts, such as Gipp's Land in Victoria, and the whole of the western part of the same colony; and New England and the Liverpool Plains in New South Wales. These districts, however, cannot be comfortably visited unless the traveller be acquainted with one or more of the resident squatters. There are small country towns scattered about, with plenty of bush inns, but the latter are not to be recommended for invalids; they are sometimes good enough for ordinary travellers, but they are all more or less rough, and quite unsuitable for delicate persons. The long distances to be travelled over the jolting roads are also trying to invalids; they are often bad enough to those who are in rude health.

I did not visit South Australia or West Australia, except touching at both colonies on our homeward journey; but I believe that Adelaide, the capital of the former, is a very pleasant town, and enjoys a good climate, except in the height of the summer months. There is a good club there, of which we were made honorary members, though we were not able to avail ourselves of the privilege. On some part of the up-country stations of South Australia the climate is said to be very fine and particularly adapted to persons with delicate lungs. I should think that no part of West Australia would be suitable for invalids.

My brother and I stayed at Hobart Town till the 18th of February, when we returned by direct steamer to Melbourne; and we left Melbourne on the 26th in the P. and O. Co.'s steamer "Bangalore" on our voyage homewards. We touched at Glenelg (the port for Adelaide) on the 28th, but we were unable to go on shore on account of the rough weather. We touched again at King George's Sound, a port of West Australia, on the 5th of March; and this brought to a close a visit of over fourteen months in the Australian Colonies.

The fixed idea which I brought away with me, and which I still most strongly retain, is that, though a visit to some of the Australian towns is a most pleasant and enjoyable experience, it is not one that can be confidently recommended to invalids, especially those with very delicate lungs, unless it can be combined with still longer visits to squatters on their stations in the country. I think, on the other hand, that life on a well-managed station in any one of the Colonies of Australia, except West Australia, well away from the sea coast, and free from malarious influences, is admirably adapted for restoring exhausted nature in whatever way it may show itself. I am satisfied that I gained great benefit by my experience of it, though my visit in the Riverina was made at about the worst time of the year. In England invalids are advised to winter on the sea-coast, which is generally less visited by extreme frost and snow than other parts of our island; but in Australia the further inland the invalid goes, the less he is exposed to the disturbing influences of the ocean, which can never be absent in such a small island as England. If an invalid can be assured of being invited to spend much of his time on one or more of these country stations, I can unhesitatingly recommend a voyage to Australia.

CHAPTER VII.

THE VOYAGE HOME.

P. and O. Steamers.—Overcrowding.—Suez.—Alexandria.—Valetta.—
Climate of Malta.—Society.—Sport.—Routes from England.—
Home Again.—Results of Voyage.

The "Bangalore" was crowded with passengers, and it was by the favour of the agent of the P. and O. Co. at Melbourne that my brother and I shared, with two other passengers, a very small cabin, about eight feet long by six feet wide, in the forward part of the steamer, instead of being suffocated in the wretchedly hot and stuffy little cabins close to the engines and boilers. I have travelled in the large steamers of the three great Mediterranean companies—the P. and O., the Messageries Maritimes, and the Austrian Lloyd's— and, on the whole, I prefer the P. and O. The food, the hours, and the general arrangements, are more suitable to English tastes; but in no steamer of either of the other companies have I ever been so miserably crowded for sleeping accommodation as in those of the P. and O. Co. In these steamers, which pass through the Red Sea and the tropics, every little cabin is fitted up with as many berths as it can possibly hold, and is often crammed to overflowing with passengers, many of whom would most assuredly die of suffocation in the tropics, did they not sleep on deck, or in the saloon, or in the passages, or indeed anywhere out of their cabins where they can get a mouthful of air. Many of the cabins are ranged on each side of the engine and boilers, and how the occupiers of them manage to live in the tropics, I never could understand.

I should recommend to invalids the journey to Australia by the P. and O. steamers, in preference to the long sea route, if only decently airy sleeping accommodation could be assured for them. The steamers are for the most part good and well found; the officers competent and educated men; the weather by no means unpleasant at the proper time of the year, which is between the beginning of October and the end of April; the route is much more interesting; and the time occupied only about half of 'that in a sailing ship round the Cape. There are generally plenty of agreeable fellow-passengers; and I have often been quite sorry to get to the end of my voyage in one of those steamers. The only drawback—but it is a grievous drawback to invalids—is the chance of being crammed into an overcrowded cabin, and poisoned with foul air every night. The fare for one berth from England to Australia in one of these crowded steamers is about the same as the fare for a whole cabin on board a sailing ship, and this must be taken into serious consideration in deciding upon the route to be selected.

There is a line of steamers lately started between England and Australia, round and touching at the Cape of Good Hope, called the "Orient Line." The steamers of this company are large, and, I am informed, well found and comfortable. I hear, however, that they also are apt to be overcrowded, and that the general arrangements are hardly comparable to those in the steamers of the P. and O. Co. The variations in temperature are also greater by the Cape route, though the steamers do not get into such low latitudes as the sailing ships are compelled to do in order to catch the westerly winds after rounding the Cape.

We had a fine and uneventful voyage to Suez, where we arrived on the 5th of April, having cool and pleasant weather in the Red Sea. At Suez we took leave of most of our fellow-passengers, as we were going to Cairo for a few days, intending to go forward by the next P. and O. steamer from Alexandria. Many of the passengers went forward in the "Bangalore" through the Suez Canal to Southampton; and others went direct to Alexandria by the mail train, and returned home by Brindisi. There is but little to see at Suez, and very little to eat at the large hotel there. We were therefore not sorry to start by the early train next morning for Cairo, where we arrived about five o'clock in the evening. I shall reserve what I have to say about Cairo till a subsequent chapter. I will only say here that my brother and I, and some of our fellow-passengers, went to the Hôtel du Nil there, and found it exceedingly comfortable. The situation is rather against it; but the accommodation and food, and the attention which we got there, were quite equal to what we met with at the great English hotel, "Shepherd's," where I afterwards passed a considerable time, and the charges are decidedly lower.

We remained at Cairo till the 12th of April, when we left for Alexandria, whence we started for Malta two days afterwards in the P. and O. Co.'s steamer, "Hindostan." We had a fine passage, and landed at Valetta on the 17th, and we remained there about a month. Valetta is a most striking-looking town: it is a veritable city of palaces, and all the principal buildings which front on the main streets formed, in past times, the splendid residences of the Knights of St. John. The hotels were all palaces, and they have a somewhat cold and gloomy appearance on first acquaintance. We went to Dunsford's Hotel in the Strada Reale, and found it comfortable in all essential particulars. Many of the bedrooms are small and dark, and the attendance in them might easily be improved; but the food is excellent, and the prices charged are most reasonable. The ordinary charge is from 8s. to 9s. a day for board and lodging, and there was an abundance of everything provided, and of excellent quality. The meat is as good as English meat, and the vegetables much better and more varied. The oranges were delicious. No other fruits were then in season, but they are plentiful at the proper times of year.

Many people think Valetta a good place in which to

spend the winter; partly on account of the goodness and reasonableness of the accommodation, and partly on account of the climate, which they say is genial and temperate from the beginning of October till the end of May. My brother and an invalid friend spent the whole winter of 1875-6 at Valetta; and, though both of them had a considerable experience of winter climates, they said they had never passed a more agreeable winter, from a climatic point of view.

There are two or three other hotels in Valetta, which are no doubt as good as Dunsford's, but I think not quite so well situated. There is also a good hotel called "The Imperial," at Sliema, a suburb of Valetta, and I should be inclined to recommend that hotel to any invalid who proposed to spend the winter at Malta. It has the disadvantage of being on the other side of the Quarantine Harbour from Valetta, but there are ferry-boats crossing every two or three minutes, which take passengers across for one halfpenny. There are good suites of apartments for families in many of the large private houses, and many English people would consider it an advantage that their own language is spoken and understood by all the principal tradesmen and inhabitants in the place.

In spite, however, of the manifest advantages of Malta, I think I should find a residence there for any length of time exceedingly dull. Riding, and even driving, are almost impossible with any comfort along the dry and dusty roads; and there is a horribly stony appearance about the whole island. The soil is said to have been all brought from Sicily, and every little patch of ground is surrounded by rough stone walls to keep it from being washed or blown away. The walks in the country are equally stony and dusty, and necessarily very limited in number and extent, and it soon became wearisome to walk up and down the streets of Valetta. There is an excellent English club there, and a good garrison library, to both of which visitors are admitted as honorary members on the recommendation of members. There is plenty of pleasant society amongst the officers of both branches of the service and their families, if an invalid be well enough to take part in it; but balls on board ship, and heavy dinners and suppers, with a walk home at night, are not considered very conducive to health for those who have to winter abroad. There are pleasant expeditions to be made from Valetta to Tunis, and to Sicily and Naples, by regular lines of steamers; and a sportsman may get good snipe shooting in Sicily; but the best ground for this is swampy and not remarkable for healthiness: he can also get good quail shooting in Malta and the neighbouring island of Gozo in the spring.

The journey from England to Malta is comfortably made in about nine days in a P. and O. steamer, calling at Gibraltar on the way. A good passage may reasonably be expected between the beginning of May and the end of October. I have crossed the Bay of Biscay repeatedly during that time of the year, and I have never experienced bad weather. To those who are haters of the sea, the voyage can be shortened by going to Naples by land, and thence in four or five days to Malta by way of Sicily. There are also steamers from Marseilles in about five days.

One of the great advantages of Malta as a health resort, is that invalids can so easily get away from it, if it be found unsuitable for them, by either going forward to Egypt and the East by P. and O. steamer: or to Sicily or Africa as above mentioned. There are much the same climatic advantages to be obtained at Gibraltar, but the accommodation and food there is greatly inferior to Valetta. On the other hand, there are a greater number of interesting excursions to be made from Gibraltar without much trouble or fatigue.

We left Malta on the 15th of May, and, after a fine passage, landed at Southampton on the 24th. I had thus been absent from England nearly two years: and I may safely say that I had made great progress towards recovery during that period. We found the usual " exceptionally " cold and late spring weather in England, and we went to Folkestone, and afterwards to Eastbourne for some weeks, before going home to the north-eastern counties. There was a bitter northeast wind blowing over the whole island, and I saw the first snow at Folkestone that I had seen for more than two years.

Invalids who have been out of England during the winter, should make a rule not to return home before the end of the first week in June, except in very exceptional seasons. May is one of the most treacherous months in the year in England, and many a man who has made steady progress during the winter and spring months abroad has lost nearly all the good he has gained, by his impatience to return home on the first appearance of summer in England; an appearance which is quite certain to be belied before he has been at home many days. Those, also, who return home at any time of the year from a long sea voyage should be most careful during their first few days on shore against catching cold, to which, for some reason or other, they appear to be extraordinarily liable. I suffered severely from my ignorance of these two rules; as I caught so bad a cold at Folkestone as seemed at first to undo all the good I had gained on my travels. This fortunately passed away after a time, and I have never since had any serious relapse. I attribute the first real improvement in my health to having taken this trip to Australia.

CHAPTER VIII.

ALGIERS.

Route from England.—"The Pearl set in Emeralds."—Hotels.—
Wine.—Weather.—Mustapha Supérieur.—Villas.—"Zammitt's."
—Climate.—Bad for Bilious Persons.—Drainage.—Want of Good
Hotel.—Marseilles.—The Bise and Mistral.—Pau.—Biarritz.
—Arcachon.—Results of Winter.

I KEPT well during the summer of 1874, but I was recommended by my doctor to go out of England for the winter. I was not ill enough to require any particular climate prescribed for me, but I was merely advised to go where more sun might be expected than at home. I had heard pleasant things of Algiers from a friend who had a house there, and I determined to try the place.

The best and shortest route to Algiers is by Paris and Marseilles, and thence by steamer across the Mediterranean. There are two good steamship companies in Marseilles, the "Messageries Maritimes" and the "Valery Frères," each of which runs two steamers weekly to Algiers in about 36 to 40 hours. These companies are considered to be equally good, and their steamers equally well found.

We went in one of the Messageries steamers, and took 48 hours in our passage, but we had contrary winds and rough weather in the Gulf of Lyons, which is a very common occurrence. We arrived off Algiers about sunset on the 3rd of November. Our first view of the place was most striking and beautiful. The old Moorish town is built on the slope of a steep hill fronting the sea, and the modern French town is laid out on the level ground between the hill and the sea. The whole town is built of light-coloured stone, and, as seen from the sea, it is perfectly dazzling in the sunlight. When the sun goes down behind the hills to the west, the town gleams white and cold as if built of marble, whilst the surrounding country is bathed in the rosy hues of the setting sun, which lights up in gorgeous colours the snowy ranges of the Atlas Mountains in the eastern horizon. The country round about the town is clothed with the greenest possible vegetation, and Algiers has thus acquired the name of "the Pearl set in Emeralds," which is fully justified by its appearance. I am not sure that this brilliant appearance is fully borne out on a more intimate acquaintance with the place. The old part of the town is most picturesque, but incredibly dirty and evil-smelling; whilst the new part, though cleaner, is far from picturesque. The new town is laid out in formal lines of boulevards, with numerous *places*, after the manner of all French towns; and all the buildings are erected in the modern French style. There is a good harbour, though rather difficult to enter in bad weather, and the steamers run quite up to the landing-stage.

There are several hotels in the town, of which the two principal ones are, the Hôtel d'Orient on the Boulevard de la République, which fronts the harbour and is near to the landing-place, and the Hôtel de la Régence in the Place du Gouvernement, a few hundred yards further in the town. The former is the largest, and generally considered the best, but the latter is quieter, and the bedrooms in front have a warmer and sunnier aspect than any of the rooms in the other hotel. Neither of them, however, can be said to be without reproach, nor to be free from the most unpleasant odours in different parts of the houses. The food and cooking are fair at both; but not very appetising, and the meat of all kinds is hard and stringy, and with very little nourishment in it. We were at the Régence, where my friend had taken us rooms, and were on the whole quite satisfied with his choice.

The other hotels in the town which are fairly well spoken of are the Hôtel de l'Europe, the Hôtel de l'Oasis, and the Hôtel de Genève; the food is said to be as good at these hotels as at the Orient or Régence, and the prices rather lower. It is most important that an invalid should secure a fair-sized and sunny bedroom at one of these hotels before his arrival, as otherwise he would probably find that the only vacant rooms are small and comfortless, with an aspect which gets no ray of sunlight throughout the whole day, and through the window of which, when opened, the air comes in anything but fresh.

Visitors read in the guide books that the pension prices of the best hotels in Algiers vary from 9s. to 12s. a day, and they are often a good deal disappointed to find on their arrival that during the winter season no pension arrangement can be made at either of the two principal hotels; there was a fixed price for the rooms according to their size and position, but everything else had to be paid for separately. The charges, however, were not unreasonable; and I should think that most visitors might count on an average expenditure of from 12s. to 14s. a day, exclusive of wine. The wine of all kinds was bad and dear. I tasted fair native wines at the private houses of some of the wine growers, but the best wines of the country do not seem to have got into the market to any appreciable extent. The wine of the country which was given us at the hotels was bad, and I think unwholesome. The same may be said for the wine which we bought under the name of Bordeaux, the greater part of which, I have very little doubt, came from Cette, which is quite near to Marseilles, and where every abomination under the sun is turned out with the name of some first-class vintage.

I think that the bad wine which we had to drink was, to some extent, the cause of the continued disorder of my liver, which almost entirely neutralised the good effects which a residence in Algiers might otherwise possibly have brought about. There is no doubt also that the bad drainage and foul smells are responsible for a great deal of sickness; and there were very few of my acquaintances in the different hotels who did not complain more or less of unusual symptoms of disorder after a prolonged residence in the town.

We stayed at the Régence till the middle of January. We had fine warm weather, interspersed with a very few cloudy and showery days, during the whole of November and until the middle of December. It then turned wet and cold till the end of the month. On fine days the sky was nearly cloudless, and the atmosphere clear and bright, and there was no relaxing feeling in the air. We had one or two very slight touches of sirocco during these two months, but nothing to make much complaint of. We began to think that we had found an earthly paradise for persons with affections of the lungs, and many of our friends were equally hopeful.

It was a particularly bad winter throughout Europe, and many persons came to Algiers from the Riviera to escape the wet and cold weather there, and were greatly pleased at the change. Our enthusiasm was not, however, to continue unalloyed. I can only speak definitely of myself, but I know that my experience was that of a great many of my friends and acquaintances. I never had any trouble with my chest, even when the cold and wet weather came on; but, after the first few weeks, I began to feel out of sorts and bilious. I got gradually worse, until my disorder culminated in a serious bilious attack early in January, which made me determine to get out of the town, and find accommodation, if possible, in the suburbs.

The principal suburb of Algiers is called Mustapha Supérieur, and it consists of numerous detached and semi-detached villas, beautifully situated in their own gardens on the slopes of the hills which rise round Algiers on the south and west sides of the town. These villas belong to the principal French inhabitants of the place, many of whom are quite willing to let them during the winter to the visitors and to reside in the town themselves. It is by far the pleasantest and healthiest way of living for visitors in Algiers to take one of these villas, and keep house for themselves; but this, of course, requires a family party, who could take one or two of their own servants with them. Good meat and vegetables are to be got in the market; and there are plenty of good cooks to be engaged; and carriages and horses, with a coachman, can be hired by the month at reasonable prices. With these conveniences, people go winter after winter to Algiers, and get a fine climate, pleasant society, and I have no doubt health, with very little trouble; and at no very great expense.

The first thing to consider is the selection of a villa; and this requires great care and caution. It is of the greatest importance that the house and garden should be as fully as possible exposed to the sun from morning till evening. The soil is rich and the vegetation luxuriant, and it is astonishing how, even in fine weather, everything looks and feels damp and chilly directly one gets into the shade, or the sun goes down. This is increased tenfold in wet weather, when the soil and vegetation get soaked, and when there is no sun and very little wind to dry it. A real sunny villa is more difficult to get than might be expected, as the residents, like all inhabitants of hot climates, have for the most part built their houses with a view to getting as much shade as possible. It is of equal importance that attention should be paid to the drainage and water supply. An English family, with whom I was acquainted, had a very nice villa in Mustapha Supérieur during the winter we were in Algiers, and they had to decamp at a moment's notice, and long before their time had expired, on account of an outbreak of low fever amongst the children, evidently caused by bad drainage. I heard a great many other complaints on the same subject. Unless these two objects of sun and good drainage be secured, I would not advise any family to take a villa in or near Algiers.

It was, of course, impossible for my brother and myself to take a villa with any chance of comfort; and when I felt that I was doing more harm than good in the town, I began to enquire about hotel accommodation at Mustapha Supérieur. I had been told of one Zammitt, who kept a kind of hotel and pension there; and I went to look at it. The house had been originally a fine old Moorish house, but so many additions had been made to it that it had almost lost its original characteristics. It was beautifully situated in a very pretty garden, in one of the best parts of

Mustapha Supérieur, and close to the high road from the town. The house was called "The Hôtel de la Ville Orientale," but it was more commonly known as "Zammitt's." Mr. Zammitt, the proprietor, was a Maltese and consequently an enterprising man, and he had made his house as comfortable as it would admit of for English visitors, of whom it was quite full during the whole of my stay there. The prices were high considering the accommodation, and I had to pay 150 francs a week for a very moderate bedroom and pension, exclusive of wine. I do not complain of the charges, as Mr. Zammitt was entitled to charge as much as he could get, and he had many more applications for rooms than he could accept; but I am surprised that no one has been enterprising enough to build a really good hotel at Mustapha Supérieur. If the accommodation were good and the prices reasonable, I think a good hotel there would certainly pay well, and it would be an enormous attraction to the place.

When I was in Algiers, reports were current that a company was being formed to buy a piece of land and build an hotel in a high and sunny situation in Mustapha Supérieur, but I have not heard that the scheme was ever carried out. Much that I have to say in disparagement of Algiers, might be obviated by the erection of a good hotel with a strict regard to sanitary details in its arrangement. Zammitt's Hotel was not nearly large enough, nor did it admit of any satisfactory alteration. The salon and salle-à-manger were good, but the bedrooms were nearly all very indifferent. The food was good and plentiful, but the attendance left much to be desired. Life was, however, much pleasanter and healthier at Zammitt's than in the town; and, until some better accommodation is provided, I recommend Zammitt's as decidedly preferable to any hotel in the town for a prolonged residence.

We had a beautiful month of January, and a moderate month of February. Indeed, during the whole time we were in Algiers, there were very few days when we could not get a walk during some part of them with moderate comfort. It is strange, however, how much even slightly bad weather is felt in Algiers; the cold wind seemed more trying, and the rain colder and wetter, and to leave a damper feeling in the air, than in any other country which I ever visited. I may here mention an instance of the ignorance of climates which prevails even amongst those who have to advise their patients where to go for winter quarters. The English doctor in Algiers told me that, in a conversation which he had with one of the first physicians in London about the place, the physician said to him that he supposed one of the difficulties which the doctors in Algiers had to contend against was the dryness of the atmosphere and the irritating nature of the desert sand drifting about the place; whereas, in fact, the soil is deep, rich, and damp, and there is no desert within 100 miles of the town.

The winter of 1874-75 was a very severe one in the south of Europe, and we heard of ice and snow all along the Riviera. The Mediterranean lost its beautiful blue colour, and nearly always looked grey and stormy. We had, however, no frost or snow in Algiers. There were three successive days of cold rain and sleet in December, but I never saw the slightest whiteness on the ground. We often, however, felt greatly starved on wet or windy days, though the thermometer would probably not be standing below 40° Fahr. out of doors.

I was rather better for my change to Zammitt's, but my liver was never right during the whole time I was in Algiers. We took one or two excursions of a few days at a time into the neighbourhood, and I generally benefited somewhat by the change; but I always relapsed within three or four days of my return to Algiers. Toward the end of March I made up my mind to leave the place, especially as my brother was also far from well in the town. In spite of heavy equinoctial gales which were blowing, and very wet weather, we left Algiers on the 20th of March, in one of the Valery steamers for Marseilles; and a terribly rough passage we had. I am thus unable to speak from personal knowledge of the weather in Algiers during the spring. There were a few beautiful days in March before we left, but more wet and stormy ones. I was afterwards told by friends who remained in Algiers till May, that it was a wet and unsettled spring there; and the believers in the place called it " a very exceptional season." I had, however, friends in Algiers, who told me that it was very similar to the preceding winter and spring, which they had also spent there; but, on the whole, not quite so cold, there having been snow visible on the ground for a few hours on more than one occasion during the preceding winter. I have since enquired of friends who were there in the winter of 1875-76, and of others who were there in 1876-77 and 1877-78, and I have heard much the same story. So far as I could ascertain, the winter we spent in Algiers was about an average season; and we had without doubt a great many warm and beautiful days during it. This I see from my diary; but I must admit that the general impression left on my mind is, that we experienced a good deal of cold and wet weather, though there were very few thoroughly wet days. The complaints of the visitors on the subject were frequent and loud; and many sarcastic pieces were written both in prose and verse descriptive of the brilliancy of the African sun, and the azure hue of the Mediterranean.

No doubt, invalid visitors at the various sanitaria in the world often expect a great deal too much in the way of climate. In Algiers, for instance, they forget that the whole rainfall for the year has to be made up during the months that they are there; and that unless it comes in sufficient quantities, the whole country would be parched up by drought in the succeeding summer. It is rather a moot point which are the rainiest months in Algiers, as they vary in different seasons; but they certainly fall between the middle of October and the end of April. Some idea may be formed of the mildness of the climate, when I mention that there were beautiful flowers to be bought in the market all through the winter; and the bouquets which were prepared for new year's gifts were as beautiful as could be bought in Covent Garden out of the conservatories in England. We had plenty of the common kinds of roses in flower in the garden at Zammitt's all the winter, and the more delicate kinds were beginning to flower freely before we left, whilst the orange trees were covered with bloom. We had green peas in profusion in January, and other vegetables were good and plentiful throughout the winter.

There was a very pleasant English society in Algiers when we were there, and I believe that this is generally the case. The entertainments were for the most part arranged for invalids, and consisted a good deal in breakfast and luncheon parties, with afternoon receptions, which often ended in a dance. I should have enjoyed my visit in Algiers thoroughly, had it not been for my liver being constantly out of order, and even as it was, I was sorry to leave the place.

There are beautiful rides, drives, and walks to be had in all directions, and visitors can hire very fair riding horses at a reasonable price. There are also numberless interesting expeditions of a few days, or even weeks, to be made in the country, without any great fatigue; and a sportsman can find game by going some distance out of the town of Algiers. The weather was, however, too unsettled during the time we were there, and it was too early in the season to admit of our making any long excursions. It is unnecessary for me to enter into details on these subjects, as all information of this kind is given fully and well in Murray's Guide Book to Algiers, which is written by a thoroughly competent gentleman, and one well acquainted with the country.

I should think that there are many kinds of chest complaints that might be alleviated by a winter in Algiers. The average temperature is a good deal higher than that of any part of the Riviera, and yet the climate does not feel relaxing, except during the prevalence of the sirocco wind, which blew occasionally whilst we were there. I cannot, however, think it is a good place for persons who have any trouble with their liver. It took me two or three years to get over the bad effects which I experienced there in a far greater degree than I ever had done before, or have done since, in any other part of the world. Two days after we landed at Marseilles, I met a gentleman whom I knew in Algiers, and who told me that he had been compelled to leave the place for the same cause, although he had gone there with the intention of spending the whole winter. Many of my friends who, on account of the arrangements which they had made, were obliged to remain in Algiers till the end of the winter season, told me that they hardly ever felt free from bilious symptoms during the whole of their residence there, and I have since met other travellers who complained of suffering there in a similar manner. Whether it was the air, the water, the drainage, or the wine which produced these bad effects, I know not;

but they probably all contributed in a greater or less degree. I have been informed on reliable authority, that during the winters of 1878-79 and 1879-80, there was a great deal of low and typhoid fever amongst the visitors in Algiers, resulting in death in some of the cases; and it is exactly what I should have expected. The drainage of the town is undoubtedly bad, as is evidenced by the foul smells both inside and outside of the houses, especially after rain; and I think that the drainage of too many of the villas in Mustapha Supérieur is very little better. French people seem to care but little for odours which would drive English people wild in a week; and, for some reason or another, they do not seem to suffer from the ill effects of such odours, or from the bad drainage which causes them, in anything like the same degree that English people would suffer.

The first and most important requirement for Algiers, is a really good hotel, well situated in a high and sunny part of Mustapha Supérieur, and built with a due regard to all proper sanitary arrangements. The winter months might well be spent there, and the spring might be most profitably and agreeably employed in visiting the numerous places of interest which are scattered about Algeria. Most of these places can be visited without much trouble or fatigue at proper seasons of the year.

When we arrived at Marseilles, we found the usual bitter wind blowing fiercely. It was the north-east wind called La Bise; but whether it be the north-east wind or the north-west wind, called the Mistral, that is blowing, the effect is much the same on invalids. So far as my experience goes, one of these winds is almost always blowing along the whole south coast of France, from Toulouse to Nice, during the spring months; and it passes my comprehension why so many hundreds of my fellow-countrymen, who are considered too delicate to winter at home, are yearly sent down to encounter them. We left Marseilles on the 23rd of March, returning homewards by way of Pau and the Pyrenees, breaking the journey at Nismes, Carcassone, and Toulouse; and we had the east wind with us all the way to Toulouse. Nismes and Carcassone are both well worth a visit, but the hotel accommodation at Carcassone is very indifferent, and hardly fitted for great invalids.

So much has been written about Pau that it is quite unnecessary for me to do more than state very shortly our experiences of the place. The town has a splendid situation on a high ridge overlooking the chain of the Pyrenees. The view of the mountains in clear winter weather reminds one of the view from the Cathedral terrace at Berne, though I think it is inferior to it in grandeur. There were, however, but few days during our visit that we could see the mountains at all. The view is best seen from the long promenade called the Place Royale, where are situated most of the grand hotels and suites of apartments, which are filled with English visitors during the winter. We went to the Hotel de la Poste, which is not on the Place Royale, but is in an open and airy situation near the Park, and is thoroughly comfortable : the food and cooking were excellent, and the charges very reasonable, which is more than can be said for some of the grander hotels. I cannot speak in similar laudatory terms of the climate. We arrived at Pau on the 26th of March, and we remained there till the 4th of May, and during that time we had fourteen wet days, three of which were thoroughly wet from morning to night : on several of these days we had also cold winds, like an English March wind, which we were told was very unusual at Pau, where the great feature of the place is the exceeding stillness of the atmosphere, and its freedom from all winds. I cannot say that I liked the place, but I was not well when I got there, and I remained out of sorts during the whole of my visit. I had friends in Pau who had spent the whole winter there, and they gave a miserable account of the weather which they had experienced ; they said that it rained almost incessantly during the months of November and December, and that it was very little better in February. January had been fine throughout, as it was over the whole of the South of Europe. There had also been some snow, and an unusual amount of cold wind during the wet months. So difficult however is it to get reliable information about the weather in different health resorts from those who go constantly to the same place and find it suit them, that I was told by one gentleman in Pau, who had been there for ten consecutive winters and had derived great benefit from the climate, that it had been a beautiful season, with very few wet or cold days. I am sure that he was perfectly honest in what he said, and that it had been a beautiful season to him. I am equally sure that the statements of my other informants were more in accordance with the facts, as they kept diaries, and showed me the records which they had entered of the weather at the time.

The country is very pretty in the immediate neighbourhood of the town, and there are charming walks, drives, and rides to be taken in every direction. There are also beautiful excursions to be made easily to the well-known health and pleasure resorts in the Pyrenees, some of which are famous for their mineral waters, which are considered efficacious in certain cases of affections of the chest, throat, and bronchial tubes. But it was too early in the season when we were at Pau, and the weather was too unsettled, to make any long excursions. There is a large English society, and an excellent club at Pau; a pack of hounds is kept up, and there is fair hunting during the winter months. There is a polo and lawn-tennis club, and a good band plays every day. I should think that Pau would be a very pleasant winter residence for a family party who were pretty well off, and who had no great invalids amongst them, but who desired to be out of England for the winter. Personally, however, I should prefer to stay at home.

We went forward on the 4th of May to Biarritz, where we stayed for about a fortnight, and then turned homewards. Biarritz is a bright, cheerful place, and we had very pleasant weather whilst we were there; it is however hardly fitted to be a health resort during the winter and spring months for people with any serious affection of the lungs, as it is too much exposed to the stormy winds of the Bay of Biscay. There is a place called Arcachon, not far from Biarritz, which is thought highly of by some persons as a winter health resort. The soil is light, sandy and dry, and the place is surrounded on the land side by large pine woods, which protect it from the wind, and which are considered to be beneficial from the aroma which they give out. There is a good-sized hotel there, and there are houses to be let in lodgings. I am told that there is now each winter an increasing number of English visitors in the place. It might well suit a family, with one or more invalid members, who wish to live quietly and inexpensively out of England during the winter months; but it looks rather dull and uninviting to the casual visitor. I am told that there is good boating and sea-fishing in fine weather, and that beautiful walks and rides are to be had in the glades of the pine forests: but of this I have no personal knowledge.

We arrived at Folkestone on the 28th of May and found the usual "exceptionally" wet and cold spring in England. I cannot say that I derived much benefit from that winter abroad. I had no trouble with my lungs, but I came home feeling generally out of sorts, and I was told by my friends that I had developed a very golden complexion, of which it took me two or three years to get rid. I had seen some most interesting places and enjoyed my winter abroad, but I cannot say that it was a complete success from a health-giving point of view.

CHAPTER IX.

EGYPT.

Routes.—Landing at Suez.—Cairo.—Shepherd's Hotel.—Climate of Cairo.—Dust.—Bad for Hemorrhage.—The Nile.—Drawbacks to Voyage.—Hiring of Dahabeah and Arrangements with Dragoman.—The "Gazelle."—The Nile Voyage.—Upper Egypt and Nubia.—Sport.—The North Wind.—Accidents to Boats.—Dragomans.—Helouan les Bains.—Luxor.—Cost of Living at Hotels and Expense of Nile Voyage.—Result of Trip.—Sicily.—Syracuse.—Palermo.—Home through Italy.

I SPENT the summer of 1875 in England and kept quite well. I was, however, recommended by my doctor to go away again for the winter, as a matter of precaution. I had a friend who was also ordered abroad, and we agreed to go together to Egypt. He proposed going to Cairo by a different route from the one which I affected, and we agreed to meet there at the end of November, and go up the Nile together in the following month.

There are three routes by which persons can get without much fatigue to Cairo. The first, and easiest to those who do not mind a sea voyage, is by P. and O. steamer from Southampton to Suez, touching at Gibraltar and Malta, and going through the Suez Canal, and from Suez by train to Cairo. This takes about fifteen or sixteen days, a fortnight to Suez, and a day thence to Cairo. The second and shortest sea voyage is through France and Italy to Brindisi, and thence by P. and O. steamer to Alexandria, and by rail to Cairo. This would take about eight days of moderate travelling: four to Brindisi, and three in the steamer to Alexandria, which is only about half-a-day's journey by rail from Cairo. The third route is by Paris to Marseilles, and thence by Messageries steamer to Alexandria in about five days. I prefer the first route, and I think that, unless a person greatly dislikes travelling by sea, it is the least fatiguing for an invalid. It is necessary to spend at least three days on board a steamer, by whichever route you go, and by the end of three days most persons get over their sea sickness, and enjoy the rest of the voyage, which may prove beneficial to many invalids. Passengers by P. and O. steamers can take as much luggage as they like without additional trouble or expense. The steamers are good, the weather in September and October is generally very pleasant, and there are nearly always agreeable fellow-passengers on their way to the East.

I started from Southampton in the P. and O. steamer *Assam* on the 4th of November. We had a good passage, fine weather, and pleasant companions, and I was quite sorry when we arrived at Suez. There is sometimes a difficulty in landing there, as the steamers occasionally anchor out in the bay about five miles from shore, and the Egyptian boatmen drive a hard bargain for landing the passengers. It is well, if on these occasions there be two or three gentlemen of the party, as the boatmen at Suez and Alexandria are much more reasonable under a display of physical force than under any sentimental feelings toward the weaker sex.

We landed at Suez about 8 o'clock in the evening of the 18th of November, and went forward next morning by train to Cairo, and took up our quarters at Shepherd's Hotel there. "Shepherd's" is *the* English hotel, and nearly all the visitors who go to Cairo with the intention of making a voyage up the Nile, go to that hotel. I have mentioned in favourable terms in a former chapter the Hôtel du Nil for families who wish to make any lengthened stay in Cairo; but persons who are desirous of making up a party for the purpose of taking a dahabeah and going up the Nile together, in order to lessen the expense of the trip, have more chance of meeting with others similarly inclined at Shepherd's than in any other hotel at Cairo. Parties are often made up in this way, but whether they turn out successfully or not is a matter of great chance: and I should never recommend such a risky expedition to an invalid.

My friend and I had each a friend who joined us, and we thus made up a party of four men, which seems to me the best possible number for comfort and enjoyment for such an expedition. I must say that I think a mixed party of men and ladies on board a dahabeah, unless members of the same family or most intimate friends, would lose half the enjoyment of the Nile voyage.

A dahabeah is a small place to live in: there is a

general saloon, and there are a greater or less number of sleeping cabins, according to the size of the boat; but they are all necessarily close together, with probably a common bath-room in the middle. It is needless to point out that, unless people are on the most intimate terms, there is every opportunity for discomfort and disagreement amongst the party, and that there must be an absence of that perfect freedom of life which makes one half of the charm of the Nile voyage.

We had beautiful weather in Cairo, and enjoyed our time there very much. We visited all the principal mosques and other places of interest in the city and suburbs, and had all the usual experiences of travellers in that most delightful place. We remained there about three weeks, which is quite long enough to see all the principal sights, and to make the necessary arrangements for the Nile voyage.

I shall not attempt to describe Cairo, nor to give any account of the wonderfully interesting monuments of antiquity which we visited during our Nile journey. These are all fully described in Murray's "Guide Book of Egypt," and in the several books of history of the Ancient and Modern Egyptians, and of narratives of Nile voyages, which are enumerated in Murray, and which every traveller should take with him in his dahabeah if he wishes to understand and appreciate what he sees on his route. I shall merely give a very short outline of our journey, with particulars of the climates we experienced in the different parts of Egypt, and the effect upon ourselves and others with whom we became acquainted during the three months which we spent on board our boat.

The climate of Cairo from the beginning of November to the end of March is very beautiful: there are sometimes a few cloudy and comparatively cold days, with some rain, between the middle of December and the middle of February, but these five months are, for the most part, bright and sunny, and not too warm. I cannot think, however, that a residence there during these months would be good for persons with affections of the lungs. The greater part of the European quarter of the city is built on what used to be a swamp, and is still, in parts, very swampy ground; and if persons are walking after sunset in the gardens of the Esbakeyah, which are in the centre of this quarter, or riding home alongside of them after an excursion into the country, they may see a fog rising stealthily all around them, whilst a cold clammy sensation, as of a wet sheet, seems to wrap them up from head to foot. This sudden change, after the heat of the day, is often productive of bad colds and sometimes fevers to those who expose themselves to it. It may be said that no invalid should be out just before or after sunset; but when excursions are made on donkeys into the desert, it is often impossible to hit off exactly the time of return home, and a single imprudence of this kind may possibly have very serious results. Moreover, the country round about Cairo is subject to the inundations of the Nile, and is thoroughly irrigated both naturally and artificially up to the edge of the desert. Here are some of the richest and heaviest crops of clover and other green and white crops that I ever saw; and the exhalations from this rich, moist land after sunset are extraordinary. There were evenings when in our boat in the neighbourhood of Cairo, and indeed for 100 miles south of Cairo, we could hardly see from one end of the boat to the other on account of the fogs. It is a common amusement in Cairo to make up parties to visit the Pyramids by moonlight, a beautiful experience, but one not to be undertaken under any circumstances by real invalids. The nights in the desert are in the winter extraordinarily cold and frosty: and this excursion entails a drive through the richest, and by consequence the foggiest, country around Cairo.

Another disadvantage of Cairo is the great quantity of sand and dust which every one must daily inhale when either walking, donkey-riding, or driving through the streets and neighbourhood. This dust, which would be bad enough for persons with chest complaints, if it were pure, is mixed up with every kind of impurity in the streets: and thus forms one of the most irritating and poisonous compounds that can well be imagined. These drawbacks are by no means absent from a journey up the Nile, but they are there in a decidedly less noxious form. I have been told by medical men who are acquainted with Egypt, that the climate both at Cairo and up the Nile, is too irritating for persons who have any great tendency to hemorrhage; and I knew of three cases who seemed to bear out this opinion during the winter we were in Egypt.

There is one other drawback to a Nile voyage, and it is a very serious one for invalids. I may as well mention it here, as I shall hereafter have nothing but praise to give to a voyage up the Nile in a dahabeah from a most interesting and health-giving point of view. This drawback is the north wind, which sometimes blows violently up the river, and is rather dangerous to boats sailing up with it, and most unpleasant and annoying to boats coming down against it. The slight danger in going up the river may be averted by care in not sailing at night when the wind is at all violent, especially in those parts of the river where the mountains close in on both or either side, and by a strict watch that the main sheet be not tied, but always held by one of the boatmen whilst sailing. But the unpleasantness in coming down the river, which is done by drifting with the current, and generally against the wind, can only be mitigated in rough weather by mooring the boat under the shelter of the river bank, and waiting for better weather. This is easy enough when the travellers have plenty of time; but when, as is commonly the case, they wish to be back in Cairo within the stipulated time for which they have hired the boat and dragoman, they often have to proceed against the high wind. In this event, it is too cold to sit on deck with pleasure, and the boat is too unsteady to read or write with any comfort in the saloon. Indeed, many persons suffer from regular sea-sickness from the continual tossing about of the boat; and one of our companions was made so miserable by it, that he preferred to come down the last 150 miles by railway. These, and the lesser troubles of mosquitos, flies, and fleas, are the only disadvantages of which I am aware in a voyage up the Nile in a well appointed dahabeah; and these lesser troubles may be reduced to very small limits with ordinary care and attention. There are no mosquitos to trouble the Nile traveller after he once gets away from Cairo. The flies and fleas are always with him, but the former are swept out of the boat two or three times each day, and the latter are checked, if not absolutely routed, by care and cleanliness in the boat and a free use of Persian powder.

I must refer my readers to Murray's Guide Book for all information respecting the hiring of the dahabeah, and the contract with the dragoman; merely saying that all the directions contained in it on both subjects, should be strictly followed. By doing this, we saved ourselves from all disagreeable questions which might otherwise have easily arisen after the voyage was over. It is not advisable to hire the boat from the dragoman whom you engage; nor let him have anything whatever to do with the hiring. Travellers should make their own arrangement with the boat-owner respecting the hiring of the boat and crew, and then make a separate contract with the dragoman for all other necessaries. If the dragoman be the owner of the boat, he will be always nervously anxious to avoid risk of any damage to it, which might possibly arise through sailing with a good breeze, or by getting on sandbanks or shoals, or in ascending or descending the cataract; and travellers will often be unreasonably hindered in their progress on this account. Delays and hindrances in sailing are, of course, advantageous to the owner of the boat, who wishes to prolong the voyage as much as possible, but they are the cause of great annoyance and irritation to the hirer.

It is, moreover, of the utmost importance that the dragoman should be made to understand from the very first that he is the servant, and not the master of his employers. He will try to assume the conduct of the expedition, and to get the upper hand in the management of the boat and crew; it is hard to withstand this if he be the owner, but if he has nothing to do with the boat, it is easy by a little firmness to put him in his right position at once.

We were a party of four, and we hired a very nice and clean dahabeah called "The Gazelle." She was a small boat, but a good sailer up the river with the wind, and a good drifter down the river against it. This last is an important point to be considered in hiring a boat. The Gazelle had a fair-sized saloon, with four sleeping-cabins and a bath-room, and I was never in a more comfortable home than during the three months we were in her. We went on board on the 11th of December, and dined and slept in our new quarters. The next day we spent on board, and we started on our expedition during the afternoon of the 14th. It is always an advantage to spend a couple of days on board the boat at Cairo before starting on the journey, so that it may be found out whether anything, and what, is wanted for the comfort of the travellers before it be too late to remedy the want. A south wind had been blowing pretty strongly for the preceding two or three days, but it shifted to the north during the afternoon of the 14th, and we managed to sail about fifteen miles before it got dark, and we came to anchor for the first night at a place called Badrashaen. We had afterwards the usual experience of fair wind, foul wind, and no wind, but I may say generally, that the wind was light or ahead during a great part of the first 100 miles, during which time the boat had to be towed along by the crew, whilst we walked alongside, very often with our guns, at some little distance from the river. During this time we had frequent night and morning fogs, and on one night some very heavy showers of rain. The days were always fine and bright and not too hot. It was not till we got past Assyoot that we lost the morning fogs. The weather then became gradually brighter, and the atmosphere clearer as we got into Upper Egypt, and this continued with very slight interruptions up to the end of our voyage. Perhaps the most beautiful clear balmy weather which we experienced, was between the first cataract and a place called Korusko in Nubia, but unfortunately there is not much of interest to detain travellers in that part of the Nile; the nights were bright and frosty, but the cold became more intense as we got further south, so that at Wady Halfeh, where is the second cataract, we felt quite starved in the saloon and cabins. The north wind was also blowing strongly and coldly during the last part of our journey, and it was unpleasantly cold even in the daytime. We did not feel the full unpleasant effect of this north wind till we turned the boat's head down the river for our homeward journey, when we suffered to a very considerable extent from both cold and the unpleasant rolling of the boat. I should never advise invalids to go higher up the river than Aboo Simbel, commonly called Ipsambool, about forty miles below Wady Halfeh; and indeed, were it not that there is a very fine and interesting rock temple at Aboo Simbel, I should advise invalids to stop at Korusko.

The north wind blew harder and colder the whole distance between Wady Halfeh and Korusko as we returned than in any other part of the river, but it was especially bad between Wady Halfeh and Aboo Simbel. There is very little to see at Wady Halfeh, and the second cataract is not nearly so fine as the first.

If I were to make another Nile voyage, I should not, for my own pleasure, go beyond the first cataract. The climate of Upper Egypt is beautiful, and nearly all the most interesting places lie between Denderah and the first cataract. Moreover, the ascent of the cataract is a most tedious and troublesome business, which occupies two days, besides the uncertain number of days during which the traveller has to wait the pleasure of the Sheikh of the cataract at Assouan.

Philæ, which is above the cataract, is well worth a visit; but that can easily be made from Assouan on camels or donkeys, a distance of only about six miles. It is a new and exciting experience to "shoot" the cataract on the return journey, but I think hardly worth the trouble and additional expense of ascending it. There is always also some slight danger of an accident to the boat both in ascending and descending the cataract, and this might prove a very serious inconvenience to the traveller, as there are no means at hand of repairing an iron boat, and nearly all the newest and best dahabeahs are now made of iron. It is probable that most travellers who have the time at their disposal, and who are making the expedition for the first time, will wish to go as far and to see as much as they can, and will make no account of the little trouble and risk of the cataract, and I hope they will feel themselves repaid for their trouble.

The scenery above Philæ is flat and uninteresting, and the desert comes down nearly to the river bank on each side. The climate of Nubia is dry and magnificent, except during the prevalence of the cold north wind; but, on the other hand, nothing can be finer than the climate of Upper Egypt, whilst the fine temple of Aboo Simbel is almost the only object of great interest between Philæ and Wady Halfeh.

The sportsman who wishes to get a crocodile must go up as high as Korusko at least, but the chances are greatly against his getting one, even if he goes up to Wady Halfeh, unless he devotes a good deal of time and trouble to the object, and has good weapons. All other game disappears above Philæ, as there is nothing for it to feed upon. It is probable that most sportsmen would be disappointed at the scarcity and wildness of the game of all kinds along the whole length of the Nile. There are thousands upon thousands of wild geese and ducks and every kind of water fowl to be seen on going up the river; but they keep at a safe distance from the dahabeahs, and can only be got at by much trouble and perseverance: and they have nearly all disappeared by the time that the traveller returns down the river, when he might be disposed to devote a little more time to them than he feels inclined to spare at the commencement of his journey. There are a few snipe in certain places, and also sand-grouse; but the latter are scarce and very wild. There are plenty of pigeons everywhere below Assouan, and the inhabitants are quite willing that they should be shot, so long as they are in the fields and not in the villages; and these are almost the only edible birds that can be counted upon until March, when the quail come down from the interior of Africa in great quantities. During the latter part of March and through April capital quail shooting can be got all the way down to Cairo, and in the immediate neighbourhood of Cairo itself.

There are, however, quantities of birds which are rare to Europeans to be met with along the whole course of the river between Cairo and Assouan, and very good collections are made each year by those who are fond of natural history and who understand the subject. But in this case, the traveller must himself be able to skin his specimens, and partly preserve the skins, or he must have some one with him who understands the business. Very few of the dragomans, or the servants provided by them, are able or willing to help in skinning the birds or preserving the skins, though not unfrequently they express themselves as perfectly able and willing to do so before the voyage is commenced. There is an admirable book, "Shelley's Birds of Egypt," which is a great addition to the library of the Nile boat.

We hired the boat and engaged our dragoman for three calendar months, and we got back to Cairo two or three days before the expiration of the appointed time, having visited at our leisure all the most important places of interest between Cairo and Wady Halfeh, and some of them both going up the river and returning.

We spent about ten days at Luxor, and we were nowhere pressed for want of time. I think that in ordinarily fair weather, three months are quite long enough to see and appreciate most of what is worth seeing in the neighbourhood of the river, but we could gladly have spent another month on our return journey between Assouan and Cairo, had we had the time to spare. We counted each day as we got nearer to Cairo as a boy counts the last days of his vacation, and we were quite sorry when a fair breeze sprung up about 150 miles from the end of our journey, and carried us swiftly homewards two or three days in advance of the stipulated time.

I referred in an earlier part of this chapter to the danger of sailing at night, especially where the mountains come down to the river. The year that we were on the Nile was the year that the terrible accident happened to the dahabeah in which the Miss Russell-Gurneys and their brother were sailing up the river. Their boat was only a few miles behind ours at the time of the accident, and we were sailing that very night with a strong north wind, and unconscious of the danger which attended it. Our dragoman always professed to be on the look-out at night, so as to stop the boat in case of too strong a wind; and, as it was in the early days of our journey, we went to bed in confidence that he would do so.

I could hardly sleep from the noise of the wind and the rushing of the water as we drove along, and at last, about 2 o'clock in the morning, I began to think it must be dangerous, as the boats are very top-heavy with the large lateen sail and raised saloon deck. I got up and went into the next cabin, and found one of my friends in the same state of anxiety, and we agreed to stop the boat. I went on deck and found everybody apparently fast asleep: the dragoman was snoring loudly in his cabin and could not be awakened, and there were only three or four dark objects visible on deck. There was one of the crew squatting at the end of the main sheet, but, so far as I could see, he was fast asleep, and had tied the rope which held it. The man at the helm was awake : indeed, he seemed to be always awake, as he steered the boat night and day. It was a dark night, and the dahabeah was driving through the water at a prodigious pace. I managed to awake one of the European servants, and to get the boat stopped. I have no doubt that we were in some peril on that night. It was a dangerous part of the river near to the mountains, and the dahabeah containing the Miss Gurneys and their brother was blown over, and the three ladies were drowned, within twenty or thirty miles of us.

As is usual in such cases, it was endeavoured to throw the blame of this accident on Mr. Gurney, who, all the dragomans said, had insisted on sailing during that night, contrary to the advice of his dragoman and captain : but I have very little doubt that no such advice was given, and that Mr. Gurney had left the matter in the hands of his dragoman and crew, just as we had done, trusting to their care and experience in Nile voyages. This sad accident cast a gloom over all the Nile travellers that year; and, I was told, sensibly diminished the number of boats which were hired during the latter part of the season. I do not think, however, that with ordinary care, there should be any danger in the journey; but it does not do to rely on the dragoman for care and attention in cases of difficulty. They seem to be all alike in being ignorant and careless of danger, when it is not very apparent, and to lose their heads and their nerves, and to be perfectly useless, when it is near and apparent. I do not think that our dragoman was worse in this respect than others, but, though he professed the greatest anxiety for our safety when we were sailing at night, and his intention of carefully protecting us against all danger, we nearly always heard him snoring peacefully in his cabin long before we went to bed ourselves; and then it was almost impossible to awake him. Eventually we gave orders to stop the boat every night about 10 o'clock; and I am sure that it is advisable to see the boat safely moored to the bank before going to bed when sailing up the river, except on very clear and calm nights.

The crews seem wonderfully clumsy in the management of their boats. It may be that European ideas would not do for the navigation of Egyptian dahabeahs, but it certainly seemed to us that, when there was the slightest difficulty, the crew set about remedying it in an exactly opposite manner to that which would be adopted by European sailors. Our crew, however, took us safely up and down the river, with very few accidents of any kind; and they were certainly a most quiet and pleasant lot of fellows. They were nearly all Nubians, and some of them very fine-looking men. We parted most amicably at the end of our voyage, and the reis, or captain, made us a speech, in which he said that he hoped we should make a second trip up the Nile in the same boat, and he would try and get together the same crew. Our dragoman was a Syrian, and a Christian, and he was, I think, quite as good as any of the other dragomans whom we met. He understood his business, and his arrangements were satisfactory. We had always plenty to eat, and of a good quality. We had, indeed, to reduce the number of dishes which he wished to give us. The mutton was excellent, as were also the turkeys and chickens. We perhaps got a little tired of pigeons, but one cannot expect any very great variety of dishes on a Nile journey.

The traveller must not look for perfection in his dragoman, and some of them are very disagreeable fellows to deal with. Truthfulness is a quality which hardly any Eastern possesses in the same degree that we expect from Europeans; and dragomans are no exception to this rule. They will tell their employers what they think the latter wish to be the fact, rather than what is the fact in reality. This should be remembered in dealing with them, and it may help to prevent the traveller losing his temper when he finds that the exact opposite of what he has been told by his dragoman is the case. Dragomans are also wonderfully ignorant of everything which the traveller finds interesting during the voyage, but they will never confess their ignorance. They are, as a rule, vain and bumptious, and the traveller must make up his mind to show that he means to be the master from the first day of the engagement, and he will thus save himself much unpleasantness afterwards. He must be prepared to complain of the very slightest deviation from the contract, even if comparatively immaterial to his comfort, as sometimes attempts will be made by very slow degrees to make encroachments, which are very easily nipped in the bud, but not so easily stopped if allowed to come to a head. Some of the dragomans are decidedly good-looking men, and very picturesque in their Eastern costumes, and many of them are much spoilt by too great notice being taken of them by the lady travellers. The best known dragomans, however, are not bad fellows, and travellers may be pretty sure that, if they make a fair arrangement with one of these men, they will have no real cause for dissatisfaction, nor will any attempt be made to extort any more money out of them, when once the bargain is struck. We were so satisfied with our dragoman in all material points, that two of our party, who went afterwards to Syria, engaged him to go with them, and they were perfectly satisfied with his arrangements during their tour.

I have made a slight digression from my observations on purely health-giving subjects, but I think that any hints respecting the safety of the boats and the comfort of the travellers are applicable to all persons, whether invalids or in good health.

We got back to Cairo on the 7th of March, all in good health, and feeling much benefited by the trip. We found cloudy weather, and even some little rain, in the immediate neighbourhood of Cairo, but, up to that time, with the exception of the showers of rain on one evening to which I have referred above, and a very light shower at Luxor on our way up the river, we had had no rain, and very little cloud, during the whole three months. We stayed ten days in Cairo on our return, and though the weather was, for the most part, beautiful and not excessively hot, we all felt relaxed and unable to make any exertion in the way of exercise. We felt this the more as, during our Nile trip, we were often out the whole day taking long donkey rides, and walking for hours through temples and tombs, or shooting under a bright sun with hardly any feeling of fatigue. Nearly all our acquaintances who had been up the river, whether invalids or in sound health, expressed the same feeling of lassitude on their return to Cairo. I cannot think that Cairo is suitable for a long residence for an invalid. He might well stay there for two or three weeks before starting on the Nile voyage, but I should recommend as short a stay there as possible on his return. I certainly would not go there with any idea of remaining the whole winter.

There is an hotel and pension establishment in the desert about fifteen miles from Cairo, called Helouan les Bains, where there are sulphur springs. We spent a few hours there on one day, and lunched at the hotel. It seemed fairly comfortable, and I should think it would be more healthy than Cairo; but the situation is very lonely, and the place must be unbearably dull.

There is also an hotel at Luxor, which has been established by Messrs. Cook since we were there. I cannot speak from personal knowledge of this hotel, but a gentleman who was there during the first two winters of its existence told me that there was ample room for improvement. He said that it was fairly comfortable during the first winter, when there were but few visitors, but not so good during the second, when there were more. Great complaints were made of invalids being put into rooms that were only just finished building, and terribly damp, and though the food was sufficient, and the charges for pension not unreasonable for Egypt, where everything is expensive, my informant told me that a disposition was shown on his second visit to make excessive charges for any small extras which the visitors required. For instance, sixpence was charged for a glass of milk, which is almost a necessity for invalids. There is plenty of rich pasture land around Luxor, and I cannot think there would be any difficulty in getting good milk at a reasonable price. We always had a plentiful supply each day during our voyage up the Nile.

Luxor is in the centre of a most interesting district for explorations, and enjoys a most beautiful climate, and can be readily reached from Cairo by Cook's steamers, which go up and down the river once a week during the winter. If the hotel were made comfortable, and the charges were reasonable, I could imagine Luxor a good place for spending some of the winter months. If, however, expense be no particular object to the invalid, and if he can make up a party to go up the Nile in a dahabeah, I should advise a Nile trip as infinitely preferable to a residence at Luxor. In the alternative, however, I should prefer Luxor to any other place in Egypt that we visited.

Egypt is, unfortunately, an expensive country to live in. The usual hotel charges in Cairo are 16s. a day for each person, exclusive of wine. This is reduced at the Hôtel du Nil to 16 francs. The charge at Cook's hotel at Luxor is, I was told, 13s. a day. But, if the traveller wishes to go up the Nile in his own dahabeah, he will find it cost from 35s. to £2 a day for each person, if the party amounts to four or five. If the party consists of a less number than four, the cost will be more; and if of a greater number, the cost will be less. This includes all charges for the hiring of the boat and crew, and the dragoman's charges for board, and all expenses of donkeys and backsheesh in visiting the various objects of interest on the journey: but it is exclusive of wine or other liquors, which have to be provided by the traveller. We sent out the greater part of our own wine from England; and, in the case of an invalid to whom good wine is important, I should advise this course to be adopted. Most travellers, however, find that they require very little wine in Egypt. Indeed, any strong wines or spirits are injurious. Good sound ordinary claret is the best, and, mixed with Nile water, or with simple lemonade made from Nile water, is the most wholesome and refreshing drink after a long day in the desert. The Nile water when filtered is excellent and wholesome.

The expense of a Nile journey can be greatly reduced if the traveller will dispense with a regular dragoman; but in that case he ought to know a little

Arabic in order to do his daily marketings; or he may hire an inferior order of dragoman to interpret for him, and to make the necessary arrangements with the cook and waiters, he himself providing his own stores and undertaking the principal management of the expedition. In the latter case, he may hire such a dragoman for about £10 a month. We saw cases of both plans tried by different persons, and they expressed themselves as satisfied with their arrangements. The plan of the regular dragoman of course relieves the traveller from all troubles of management.

An expedition can be made from Cairo to the second cataract and back in one of Cook's steamers, but that is more for a sound person than an invalid. The advantage to a person in health may be the saving of time; whilst an invalid wants to linger for three months at least in the enjoyment of the beautiful climate of Egypt.

The life in a dahabeah is especially adapted to those who are strong enough to be out for several hours in the day riding on donkeys and visiting the wonderful temples and tombs of antiquity, which are for the most part situated just in the desert, and beyond the range of the inundations of the river, and to those who can take plenty of walking exercise in the sunshine. But it is also adapted to the less robust invalids, who can sit or lie about on the deck of their boat all day, and nearly every day, going up the river, whilst the boat is ever moving on through new scenery, and they have all their comforts around them, without the nuisance of perpetually packing and unpacking their boxes. These less robust invalids have to be careful not to expose themselves to the cold north wind when it blows strongly coming down the river; and all invalids ought to turn into their saloon a little before sunset, when a great change in the temperature takes place, even if some of them are able to turn out again afterwards on clear bright nights.

Certainly we all greatly benefited by our Nile voyage. One of our party had had a very serious attack on his lungs during the preceding winter in England; he was fairly well when we started, but he could not be called strong. He returned to Cairo quite well; he then went to Palestine for five or six weeks, undergoing some hard travelling there, and returned to England by way of Constantinople and Athens; and he has never since required to winter abroad, and he is, to all appearances, as strong as he was before his attack. Another of our party, who was constitutionally delicate, seemed decidedly stronger and better for the trip, and though he has since spent winters abroad for the sake of precaution, he is now fairly strong, and he hopes to be able to remain at home in the future. The third of our party started out and returned in a perfectly robust state of health. I, the fourth, returned to Cairo in good health and much stronger than when I started, and I have never lost the good effects of the Nile voyage. Two of our party went to Palestine, but the other of my friends and I thought it advisable to avoid the roughing which is necessary during some part of that expedition, especially as it was early in the season, and we had heard bad accounts of the weather in Palestine during that spring.

We returned to Suez, and went forward by P. and O. steamer to Malta, where we arrived on the 24th of March. We remained there for three or four days, and then started for Sicily in one of the steamers of the Florio Co., and we arrived at Syracuse very early on the following morning.

Syracuse is a miserable town, but it contains some most interesting remains of its former greatness in its ruined amphitheatre and theatre, and the stone quarries in which so many of the best born youths of Athens worked and died after the ill-fated expedition against the city more than 400 years before the Christian era. All these wonderful ruins can be visited in five or six hours; and there is a train which leaves Syracuse for Messina in the afternoon, which obviates the necessity of spending the night at Syracuse. We went forward by this train to Catania, and on the next day to Messina, where we rejoined our steamer, which had gone round the coast, and we went forward in her to Palermo, where we remained about ten days.

Palermo is a very interesting town, and most beautifully situated, facing the Bay of Palermo, on the north-west corner of the island. The climate in the winter is generally considered to be very soft and mild, and the beautiful flowers which were blooming in profusion when we arrived there seemed to bear out that view. It was, however, cold during our visit; the weather was fine, but there was a bitter north wind blowing in from the sea during the whole time we were there. I suffered greatly from rheumatism, though I had no signs of it before we got there, nor after we left the place. The air was far from relaxing, but the climate did not suit either my friend or myself. I seemed to have lost all inclination or power to walk about the streets, and I always felt completely exhausted after a very few hours of sight-seeing. I found an English friend in Palermo who had spent the winter there, and who was just recovering from an attack of jaundice, which he had contracted in the place; he was a medical man, and he told me that he considered the place was unsuitable for persons who had troublesome livers; and I have no doubt that that was the cause why the place did not suit me. My friend told me that the weather had been cold and wet during the greatest part of the winter, and that he had not formed a favourable opinion of the place as a winter residence for invalids. It is, however, impossible for anyone not to feel a great attraction to Palermo. The town is fine, and it contains many objects of very great interest to travellers. The country round about is very beautiful, but it was not safe when we were there to go far beyond the walls of the town, on account of the brigands, and the same cause prevented excursions being made to the most beautiful and interesting districts in the interior of the island. There had not been many English visitors in Palermo during the winter preceding our visit, but I am informed that an increasing number go there each year, and that the place is highly recommended by some of the principal London physicians in cases of chest complaints. There are several hotels of more or less prepossessing appearance, and many houses to be let out in lodgings or occupied as pensions. We were at the Trinacria Hotel, which faces the bay and is comfortable; but I should not care to remain there for the whole winter. I found ten days quite long enough to see all that it was possible to see of the town and neighbourhood.

We left Palermo on the 10th of April for Naples in the same steamer in which we went there, and we arrived at Naples early on the following morning. We eventually returned home by way of Rome, Florence, and Turin. I am not going to attempt to describe any of these beautiful Italian towns. They will be well known to most of my readers, and to those who know them not, and who are in sound health, I can only say, go and see them. But in a book of this kind, which is almost entirely intended for invalids, I must reluctantly add that they must be visited with the very greatest caution. The time of year to be chosen, the situation of the hotel or apartment in which to reside, and what to eat, drink, and avoid, are all subjects of the greatest importance to the invalid, and indeed to all who visit these towns. In no sense can they be included amongst the sanatoria of the world.

CHAPTER X.

CAPE OF GOOD HOPE.

St. Leonards.—Voyage to the Cape.—Cape Town.—Hotels.—Drainage and Water Supply.—The "Cape Doctor."—Wynberg.—"Cogills" and "Rathfelders."—Scenery.—Constantia.—Vines and Wine Making.—Exorbitant Duties.—The Erste River.—Ostriches.—Railways.—Worcester.—Sport.—Weather.—Climate for Invalids.—Bloemfontein.—Methods of getting to it.—Ox Waggon and Gear.—"Trecking."—Expense of and Requisites for Journey.—Results.

I REMAINED in England during the whole of the rest of the year 1876, feeling perfectly well, and able to get about almost as well as before my illness. I spent the months of November and December of that year and of January, 1877, at St. Leonards, and I found the place suit me. I have been there twice in the winter since that time, and I have had no reason to alter my opinion of the advantages of the place as a winter health resort for invalids. It faces nearly due south, and has a very extended frontage to the sea, and has thus the benefit of every ray of sunshine of which an English winter admits; and it is protected by the hill at the back from the north winds. It cannot perhaps be called a bracing place, but the air seems more invigorating than that of Torquay or Bournemouth, whilst less keen than that of any other watering-place on the south-east coast. The healthiest part of St. Leonards is said to be the high ground at the back of the town, but it has the disadvantage of requiring fairly good lungs and walking powers to get to it. It is proved by carefully obtained statistics, that Hastings and St. Leonards show the highest percentage of improvements in cases of affections of the lungs of any of the health resorts of England. The country in the immediate neighbourhood of the town is exceedingly pretty, and there are beautiful walks, rides, and drives to be had in all directions. There are packs of fox-hounds and harriers at no great distance for those who are strong enough to enjoy an occasional gallop, and the nature of the country is such that a good deal of sport may be seen with much less fatigue than is necessary in better hunting districts. To those who are unable to take hard exercise, the long parade, exposed to the sun during the whole day, and dotted with resting-places throughout the whole length of it, affords a splendid promenade. There seems to be an almost inexhaustible supply of good houses to be let furnished, or in apartments, at not unreasonable prices; and the shops both in Hastings and St. Leonards are good and well supplied with nearly everything that can be wanted by invalids or persons in good health. The east winds are trying in March, but not more so, I think, than in most other parts of England. I spent the whole of the winter of 1880-81 at St. Leonards, from the end of October to the middle of April; and though it was an exceptionally disagreeable winter there, I am satisfied, from the weather reports which we received from all quarters, that we were better off than in most other parts of England. I can speak well of the place as far as my own experience goes; and it has answered well as winter quarters for many of my friends. I had been advised by my doctor to leave England before the spring of 1877, and I decided to make a voyage to the Cape of Good Hope.

There are two lines of mail steamers between England and the Cape, " Messrs. Donald Currie and Co.," and the " Union Co." They are, I believe, equally good : but as I went to the Cape and returned to England in steamers of the former line, I cannot speak from personal experience of the latter.

The *Dunrobin Castle* in which I went out, and the *Balmoral Castle* in which I returned, are both good steamers, and were then quite new and built on the same principle. Each has a very fine saloon the whole breadth of the ship, well lighted and ventilated, and with no cabins opening into it. The advantages of this arrangement are that in bad weather the passengers have a really comfortable room in which to sit, and that they are not annoyed by the horrid noises and odours which generally proceed out of the cabins in such weather; but the disadvantages are, that there are comparatively few good cabins for the size of the ship, and that those aft of the saloon are dark and too near the screw to be comfortable. There are a few deck cabins on all the new mail steamers, and they are much more airy and comfortable than the majority of the cabins below, and, except in very bad weather, better suited for invalids.

I left St. Leonards on the 22nd of January, and went on board our steamer, which was lying in the London Docks, the same evening, and we started early next morning for Dartmouth, where we took up those of the passengers who wished to avoid the Channel. We left Dartmouth on the 27th, and we arrived at Capetown on the 17th of February, having made the fastest passage then on record, a very few hours over twenty-one days. This run has been often beaten since by the new steamers of both companies, and it may now be taken as a good average passage. The steamers in which we went and returned were fairly well found in provisions, but great complaints were made of the wines on both ships. The claret was certainly almost undrinkable; and there was a very insufficient supply of ice for the hot weather. We had fine weather on both voyages, without meeting with anything worth mentioning in a book of this kind.

The advantage of the voyage to the Cape over the voyage to Australia is that the changes of temperature are not nearly so great. Within two days after leaving Dartmouth the steamer gets into warmer weather; and if the passage be a fair one, and made at the proper time of year, the passengers ought not to suffer from cold during the rest of the voyage. There is generally a pleasant break of a few hours at Madeira, and some of the steamers touch at the islands of Ascension and St. Helena. We landed at Cape Town on the 17th of February, and went forward on the same day by train to Wynberg, having been recommended to stay at Cogill's Hotel there.

The view of Cape Town from Table Bay is rather striking, but not so fine as I had expected. The town is built on a narrow strip of land between Table Mountain and Table Bay, the back of the town gradually creeping up the base of the mountain. It cannot be called a beautiful or a pleasant town, though there are some rather fine buildings scattered about in different parts of it. The streets and shops are fair, and there are some pleasant-looking houses and gardens on the rising ground at the back of the town, but there is a scorched and dusty look about the place at first sight, which is fully borne out on a closer acquaintance with it. There are two or three moderate looking hotels in the different streets, but they have all the same scorched and dusty look about them, and we were glad that we were not obliged to take up our quarters at any of them.

Cape Town is not a desirable place for persons with delicate lungs. It is subject to dust storms, sometimes violent enough to hurl pebbles and small stones through the air. These storms are raised by the north-east wind, which is there considered the general scavenger of the place and purifier of the air, and is known by the name of the "Cape doctor." The water supply is very inadequate, and so bad is the drainage, and so numerous and noisome the smells, that the place would be almost uninhabitable without this wind. It often rises about 1 or 2 o'clock in the afternoon, and rages fiercely for two or three hours, and subsides before sunset. It has its uses, but one is apt to forget them if obliged to be out in the height of it. Even with the assistance of the "Cape doctor" typhoid and scarlet fevers are by no means unknown in the place, and small pox is said to be prevalent, especially amongst the Malay portion of the population. During the time that we were at Cape Town there was a regular epidemic of scarlet fever, which extended even as far as Wynberg, sparing neither rich nor poor. The dust storms are fortunately almost confined to Cape Town, and rarely extend even as far as Wynberg, which is only distant about eight miles from the town—but even at Wynberg the dust is sometimes very trying; and, as the roads are formed in the light red sand of the district, the traveller is apt to find, after a short stay there, that his clothes have all taken the same ruddy hue. This cannot be very good for persons with delicate lungs.

The short line of railway between Cape Town and Wynberg passes under and round Table Mountain, and through the pretty suburbs of Rondibosch and Clairmont. Wynberg is the most distant suburb from Cape Town, and is charmingly situated in the middle of a pine forest, and consists for the most part of numerous detached residences, with pretty gardens fronting the high road from Cape Town to Simon's Bay.

Cogill's Hotel is situated about the middle of the place, and within five minutes' walk of the railway station, and we found it thoroughly comfortable and homely. It is not a large building, and the traveller must not look for the grandeur and magnificence of a continental hotel; but it is quiet, clean, and comfortable, and the food plentiful and good. I was told that it was by far the best hotel in South Africa; and certainly, we found nothing at all comparable to it during our short stay in the Cape Colony. I am informed that Mr. Cogill has made his fortune, and has now retired from the management of the hotel, but that it is still called by its old name of "Cogill's." There is another hotel, called "Rathfelder's," about two miles beyond "Cogill's" on the same road. The external appearance is a little rough, and I was told that the internal arrangements left something to be desired: but I heard it fairly well spoken of by friends who were staying there; and if I were unable to get a room at Cogill's, which is often quite full, I should decidedly prefer going to Rathfelder's to remaining in Cape Town.

We stayed at Wynberg from the 17th of February till the 24th of April, making short excursions thence for a few days at a time to neighbouring places; and during that time we had for the most part beautiful weather until about the last week or ten days of our visit, when the weather began to break up. From what I could learn from the residents, the months of February, March, and April, are generally very pleasant; the extreme heat of the summer is past, and the wet season, which is the late autumn and early winter, has not commenced. Moreover, it is the height of the grape season; and the grapes at the Cape are most excellent and plentiful, and very wholesome. Many of our acquaintances had vineyards, into which they allowed us to go at any time and eat as many bunches as we liked; and nearer at hand were vineyards where, on payment of a shilling or two, we might do likewise. The grapes are as good as most English hot-house grapes, and they seem to be lighter and easier of digestion from growing in the open air; and they are looked upon in the colony as a sovereign remedy for nearly every kind of complaint. The residents pride themselves also on the excellence of many of their other fruits, especially melons, peaches, and pears; the melons were delicious, but we were rather too late in the season for the peaches, and too early for the pears. The beef, mutton, and poultry at Cogill's were all good, and the vegetables excellent. Good milk could also be got without difficulty, and the prices for everything were reasonable.

The country in the immediate neighbourhood of Wynberg is very beautiful. The whole district is studded with charming country houses of the inhabitants of Cape Town, which are surrounded by beautiful gardens and richly-wooded grounds and vineyards. There are hundreds of acres of pine woods, from the outskirts of which sandy plains covered with heath and scrub extend for miles into the interior of the country. When the heaths are in bloom these plains are carpeted with flowers of every varied hue, and look like one vast garden as far as the eye can see. We were too early in the year for most of the bulbous flowers, for which the Cape Colony is so renowned, but their gorgeous colours must greatly add to the beauty of the scenery at the proper season. The ground rises steeply at the back of Wynberg, and the sloping hillsides are clothed with unfamiliar trees and flowering shrubs of various kinds and great luxuriance. On these slopes grow in profusion the singularly beautiful trees there called silver trees, the leaves of which, when stirred by the breeze, glisten in the sun like burnished silver. The silver tree, of which I do not know the proper botanic name, is peculiar to this one spot of ground, and is not found elsewhere in the Cape Colony, nor, to the best of my belief, in any other part of the world. Behind these slopes rises Table Mountain, with the strangely flattened summit from which it derives its name, and which makes a fitting background to a picture which requires to be seen to be fully appreciated. The walks, rides, and drives in the neighbourhood of Wynberg are beautiful, though somewhat limited in number, and fair horses can be hired by the week or month in Cape Town, and kept by the visitor at his hotel. Nearly all exercise is taken on horseback, as the climate is not favourable to long expeditions on foot.

Wynberg is a charming place in which to spend two or three of the best months of the year, but I do not think it is suitable for a prolonged residence for an invalid with delicate lungs. The variations of temperature are great and sudden, and a bright hot morning is apt to be succeeded by a cloudy afternoon, with a high and cold wind raising clouds of dust in all directions. The vegetation is so luxuriant that after a few hours of rain the atmosphere seems to be suffused with moisture in spite of the sandy soil, and everything that one touches feels damp and clammy. These variations are no doubt attributable in a great measure to the immediate proximity of the sea on both sides of the Cape Promontory, in a latitude which is especially subject to frequent and heavy storms; and the further inland the traveller goes the less frequent are these changes and the more equable is the climate.

When we were at Wynberg, we paid many visits to the far-famed vineyards of Constantia, which lie within a very short distance of the place. The proprietor of the principal vineyard called High Constantia is Mr. Henry Cloete, whose name is almost as well known in England as at the Cape, he being one of the largest exporters of the celebrated Constantia wines, and noted for his hospitality to all English visitors in the colony. We tried many of the wines made at Constantia, and some of them were very good, but far too sweet for English tastes. We were told that the soil of the district was so rich and the climate so fine that all the vines, no matter of what description, got the rich Constantia flavour after a few years; and that this accounted for the sweetness in all the best Cape wines; and that this sweetness could only be obviated by checking the fermentation in the process of making the wine, and thus rendering it unfit to keep for any length of time and unsuitable for exportation. We tasted some claret made at Constantia from vines imported by Mr. Cloete from the Lafitte vineyards at Bordeaux, but it had completely lost the Lafitte flavour, and tasted more like a mixture of claret and port than anything else. The very grapes on the vines had become large and sweet in their new home. We were on the other hand told that, if the grapes were more carefully picked and sorted, and if the wine were more skilfully made and then allowed to stand for at least three years in cask before bottling, it would lose its sweetness, and would become delicious wine; and that the fault of the South African wines lies in the fact that none of these conditions are commonly fulfilled. My own opinion inclines to this view of the case. I can, at any rate, say that the only drinkable native wines that we tasted in the colony were at the private houses of the vine growers and makers of the wine, where it had no doubt been more carefully made and kept than the wine intended for sale; and that all the colonial wines we bought and tried at any of the hotels or restaurants, under whatever names they were described in the wine list, were unpalatable, and I think unwholesome. Most of the residents drink French and Spanish wines and English beer and porter; but it seems a great pity that this should be necessary in a country which appears to have all the most favourable conditions for making excellent wine both for home consumption and exportation. We were told by one of the large vine growers in the colony that it pays him better to make common wine which he could sell at a low price than to take more trouble and make a higher class wine, for which it was doubtful whether there would be any sufficient market. I tasted this ordinary wine, and it seemed to me perfectly undrinkable; but there is a considerable consumption of it amongst the poorer classes in the colony. The result of this system is that Cape wines have got such a bad name that even the better kinds have no sale amongst the well-to-do residents; and they are looked down upon with undisguised contempt in England.

There is a great want of energy amongst the inhabitants of the Cape Colony, English as well as Dutch. Almost everything that is wanted for use or ornament is imported, and the revenue is made up to the Government by the imposition of exorbitant duties upon everything which comes into the country. All manufactured goods and implements of husbandry cost very high prices in the Cape Colony by reason of these duties; and trade and agriculture are hampered accordingly. If a resident gets new clothes or furniture from England, he has to declare the cost price to the revenue officer: ten per cent. is then added to the price as the estimated value of the article in the colony; and an extraordinary high rate of duty is charged upon this estimated value. The traveller who visits the colony for the sake of sport has to pay an exorbitant duty for each gun and pistol that he takes with him, besides being pestered by an incredible number of formalities before he can get the requisite Government licence for carrying arms. I am in the habit of carrying a small revolver on my travels; it was safely packed at the bottom of my portmanteau, and I declared it in reply to the question whether I had any firearms. I explained the size of it, but the revenue officer insisted on my unpacking my box and producing it for inspection. Some papers were made out, and I had to pay one pound for the duty, though the revolver was only worth about thirty shillings. On my remonstrating at this high rate of duty, I was told that I was lucky in not having to pay ten shillings a barrel, or give up my pistol.

There can be little doubt that the great depression in trade which existed when we were at Cape Town, was in a great measure attributable to these high duties. There are plenty of resources in the colony, if they were only properly utilised; and to an unprejudiced traveller, there seems no reason why South Africa should not keep pace with Australia, which has hitherto quite outstripped her in the race for wealth and position.

The country around Cape Town, except in the direction of Wynberg, is flat and uninteresting; the land is poor and stony and covered with low scrub for many miles, but it becomes much prettier and more cultivated when one gets near the foot of the mountain ranges beyond the Erste river. We paid a visit of a few days to a gentleman who kindly invited us to see his farm on the Erste river.

In South Africa, the landowners are called farmers, and their estates farms, in contradistinction to the squatters and stations in Australia. Our host had a large vineyard, and was a maker of wine on a considerable scale. It was just the time of the gathering of the grapes, and the commencement of the wine-making season; and we had thus an opportunity of seeing all the processes. The grapes are still trodden out by Kaffirs, some of whom were splendid-looking fellows, and able to go on for an indefinite period at what seemed very like hard work on the treadmill, singing and laughing merrily all the time. They seemed steady hard-working men during the work hours, but we heard many stories of their drunkenness and quarrelsomeness when off work, especially on Sundays, after they have got their week's wages. There is a certain spirit made out of the refuse of the grapes, and out of the dregs and lees of the wine casks, which is commonly known as "Cape Smoke." It is cheap and fortunately very nasty, so that only the most confirmed drinkers will have anything to do with it; but when they once get into the habit of drinking it, nothing else is strong enough for them. It tastes like a mixture of spirits of wine, petroleum, molasses, and cayenne pepper; and as the test of the goodness of the spirit is to a Kaffir whether it burns his throat like a hot coal and makes him choke for several minutes afterwards, he finds that Cape Smoke satisfactorily fulfils the required conditions. We tasted two or three different kinds of wine, all made by our host and kept for many years in his cellars. They were all excellent of their kind, and would have well borne exportation, and would, I think, have found a fair sale in England; but our host told us that it paid him better to make a greatly inferior wine, for which there was a ready sale at a low price in the colony. It requires a much larger capital than the farmers in the Cape Colony are at present able or willing to invest in the business, in order to make and keep for a sufficient length of time a large stock of really good wine; but if this capital were invested, I see no reason why the Cape Colony should not produce some of the finest wine in the world.

Our host had also an ostrich farm at Erste river, and we had thus an opportunity of seeing something of these birds in their domesticated condition, and of learning how profitable they are to their owners, when properly managed, and in a district which suits them. The birds had been lately plucked when we saw them, and they presented a most ridiculous appearance, stalking about in their nakedness, resembling in a remarkable manner a group of ballet dancers on the stage of a London theatre. There are different methods practised by different owners of taking the feathers at the proper season. Some pluck them out bodily, whilst others consider that process somewhat cruel, and cut off the feathers close to the skin, when the stumps fall out at the usual moulting time. We were assured by gentlemen who practised both methods, that the birds did not appear to suffer anything like the amount of pain under either system that might be expected; they mope and are shy of feeding for a day or two after the operation; but after that time, they feed and go about as usual. Ostrich farming has now become a very important branch of industry in the Cape Colony, and as it requires no great amount of technical knowledge, if only a suitable country be selected, it would afford an excellent occupation for any young Englishman who is possessed of some little capital, and who is constitutionally delicate and in want of work in a country with a good climate. He must not expect all the comforts and conveniences of an English home, but he could easily have all necessaries about him, and he could spend nearly all his time out in the open air; and, if moderately successful, obtain a very comfortable competence. There is a good deal of useful information on this subject in Mr. Anthony Trollope's book on South Africa.

We made another excursion to a town called Worcester, which was then the terminus of the main line of railway, which is intended, in course of time, to traverse the entire colony from north to south. The line is now opened as far as Beaufort, some 200 miles beyond Worcester. There is a great difference of opinion in the colony as to the policy of extending this line of railway further northwards. The inhabitants of the eastern provinces affirm that the only sensible way to open up the country, is to run lines of railway from the several ports on the coast into the interior, extending them from time to time as necessity requires, whilst the contemplated line will run, they say, for some hundreds of miles through a howling wilderness, which is incapable of being brought under cultivation, and which will lead nowhere. The dwellers in and about Cape Town, on the other hand, say that the proposal of their neighbours would no doubt suit them very well, and would benefit all the towns on the coast, but it would ruin Cape Town; and they assert that, though the contemplated line may not be remunerative for some time to come, it will ultimately open out a much larger area of at present unproductive land, and thus prove more beneficial to the colony. The latter view has at present prevailed; but it would seem to an unprejudiced person, that there is much to be said in favour of the former view. The principal argument against the suggested coast railways is, that the harbours along the east coast are all bad, and that it would require an enormous expenditure to make them available in all weather for ships of any tonnage, and that it is doubtful whether this could be satisfactorily done at any cost. These are questions that will have to be settled in the future, if the colony is to prosper as a whole, irrespective of the interests of any particular part of it.

Worcester is a clean and pleasant town, with several decent looking inns. We went to the "Central Hotel," kept by one Perkins; and we found it sufficiently comfortable. I should think that Worcester is a far more healthy place of residence than the neighbourhood of Cape Town, though there are swamps in the immediate neighbourhood of the town, suggestive of damps and mosquitoes in the right season. There is fair snipe shooting to be got in the swamps by those who are strong enough for rough walking amongst prickly reeds, and in water sometimes up to the knees. There are also partridges, and an occasional buck to be met with in the neighbourhood of Worcester, but it requires hard walking to get near them. I was disappointed to find game of all kinds so scarce in the parts of the country that I visited. I never saw a single head of any of the numerous kinds of buck with which the colony used to abound, and but very few partridges. There are plenty of buck to be found in the less frequented parts of the country, but they are very much scarcer now than they were a few years ago. There are still large herds to be met with in the Transvaal and Orange Free State; but even there the numbers are sensibly diminishing. They were killed in thousands by the Boers in past years as vermin, and for the sake of their skins; and the stock has never recovered from the wholesale slaughter. There are plenty of quail to be got in the neighbourhood of Cape Town in the proper season; but, with this exception, the sportsman who goes to Cape Town, in the hope of sport, will be woefully disappointed.

The fine weather which we had enjoyed for the greater part of our stay at the Cape began to break up about the middle of April; and though we had one or two fine days afterwards, we had a far greater number of rainy and windy days, with a regular storm on the last day of our visit. The atmosphere, which had up to this time been light and clear, became extraordinarily charged with moisture, and the clothes in our bedrooms felt damp and clammy. We heard afterwards that this damp weather continued for several weeks, and that the late autumn was unusually wet and stormy.

An acquaintance of ours, a medical man, who, with his wife, had gone out in the same steamer with us, and who had taken and furnished a house in Wynberg, with the intention of residing there on account of the delicate state of both of their lungs, left the place within a few months of their arrival, in disgust at the climate, and returned to England. This confirmed the opinion which I had previously formed, that the neighbourhood of Cape Town is not a suitable place for any long residence for an invalid with a delicate chest.

I was perfectly well during the two months we spent at Wynberg, but then I was in very good health when I went there, and was merely there to avoid the English spring. My friend, who was in much the same general condition as myself, found the place unsuitable to him, as he suffered there so much from asthma, which was an uncommon complaint with him. He remained away on our expeditions into the country longer than I did, as he got rid of his asthma directly he got well away from Cape Town. He found Worcester suit him, and also one or two of the small towns in that neighbourhood, but he always had a recurrence of the complaint on his return to Wynberg. I should say that if an invalid were to land at Cape Town either in the winter months of June, July, or August, or in the late spring or early autumn, he might safely, and perhaps profitably, spend a month or two at Wynberg; but I think that, unless he be prepared to spend some months in South Africa in the way in which I am next going to touch upon, or unless he be particularly fond of a sea voyage, he might easily spend four or five months with greater pleasure and advantage to himself than by making an expedition to Cape Town.

I had neither the time nor inclination to try what I have been told by many invalid acquaintances, and by doctors on the spot, is the great remedial measure to be adopted by an invalid in South Africa who has the time and money at his disposal, namely, an expedition in a wagon in the interior of the country. I have obtained my information on this subject almost entirely from two or three friends who had made these expeditions solely for the benefit of their health; and they all agreed in saying that they had derived very great benefit from them. The life is rough and rather monotonous, but it is spent entirely in the open air, and, if the proper season be chosen, under a clear sky and in a brilliant atmosphere. I have the same belief in the great advantage of getting away from the sea-coast in South Africa as I have in Australia, and an absolute faith in the sun and dry fresh air. These things may be combined almost to a certainty in a wagon expedition, called in South Africa "treck-ing," if only the proper conditions be fulfilled. I will endeavour to state as shortly as I can what these conditions are.

Bloemfontein, the capital of the Orange Free State, is one of the best starting places for such an expedition, and as it has also a considerable reputation as a sanatorium for persons with delicate lungs, I will here explain the different methods by which a traveller can get there from Cape Town. Nearly all the mail steamers from England to Cape Town go forward to Port Elizabeth, otherwise called Algoa Bay, which is the principal port of the eastern province of the Cape Colony. The steamers generally remain at Cape Town three or four days discharging cargo, and a traveller would thus have an opportunity of seeing the place and neighbourhood, and going forward in the same steamer to Port Elizabeth. It is about two days' journey from Cape Town to Port Elizabeth, and the voyage is often a very rough one. There are steamers of both Donald Currie and Co. and the Union Co. which go direct from England to Port Elizabeth, and if the traveller does not care to see Cape Town, it would be easier and cheaper for him to go by one of them.

There is a moderate inn and a fair club at Port Elizabeth, but it is not a place in which I should advise an invalid to make any lengthened stay. There is a railway from Port Elizabeth to Graham's Town, a distance of about seventy miles. It was not opened the whole way when I was at the Cape, but a mail cart met the train at a place called Sandy Flat, and carried the passengers into Graham's Town; but this railway has now probably been completed. Graham's Town is a pretty and healthy town, lying about 1700 feet above the sea, and there is a fair hotel there, which might be made very good with a little more attention to cleanliness and civility.

The distance from Graham's Town to Bloemfontein is nearly 400 miles, and it would be a very great advantage to an invalid if he could fit himself out at Graham's Town, and make the whole journey in his own wagon. I am informed, however, by a friend who has had considerable experience in wagon travelling in South Africa, that this course would have great practical difficulties by reason of the difference of the herbage of the comparative lowland country for some distance after leaving Graham's Town and the herbage of the high plateaux of the Orange Free State. The difference is so great that oxen which have been accustomed to one kind of herbage would almost certainly suffer, and might probably die, from the effects of the change before the end of the journey; and even if they were brought to the end by skilful management, they would be almost unsaleable in Bloemfontein from the bad condition in which they would certainly arrive, and from the great probability that they would do no good in their new country.

There are three other methods by which the journey may be performed. First, by Cobb's coach, which starts weekly from Graham's Town, and takes five days on the road. This is a terrible journey, and one most trying to an invalid; the roads are steep and rough, and in wet weather sometimes almost impassable. The passengers are not unfrequently requested to get out and walk through a quagmire when the night is coming on, as the only chance of arriving at their destination for the night. When the weather is fine and hot, the misery of the passengers is almost as great. The food is bad at nearly all the stopping-places, and the sleeping accommodation very indifferent. It is possible to get out and stretch one's legs occasionally by walking when the journey is up hill, but that is the only relaxation during the whole five days of travelling; and this can only be done by people moderately sound in wind and limb. This is by far the cheapest and most usual method of making the journey; and I believe that it is not a common occurrence for people to die during the operation; indeed, one gentleman in Cape Town, who is troubled with a torpid liver, told me that he absolutely derived benefit from the continuous jolting on the journey; but I fancy that his must have been an exceptional case. A second method is for the traveller to make an arrangement with a transport rider to carry him, bag and baggage, the whole journey. This is performed in an ox wagon in about a month or five weeks; and the difficulty respecting the change of herbage to which I before referred is got over by the spans of oxen being only taken half way, and the loads being transferred to other wagons, in which the remainder of the journey is performed. This may be a fairly comfortable journey, but it is dependent, of course, in a great measure, on the man with whom the bargain has been made. It is not an expensive way of travelling, and there is no responsibility; and a good deal of experience of wagon travelling may be picked up on the road. The traveller may usefully occupy his time in learning "Cape Dutch," which is almost indispensable for persons travelling in their own wagons in the Transvaal and Orange Free State. The third and most comfortable method is to buy a Cape cart, or even a light spring wagon and four horses at Graham's Town, and drive the whole way to Bloemfontein, the traveller taking his own time and staying at different places on the road. This would, under ordinary circumstances, take about ten days or a fortnight. It is not so expensive a plan as may appear at first sight, as the cart and horses, if properly chosen and cared for on the journey, would be saleable at Bloemfontein for at least two-thirds of the original price.

Bloemfontein is a comparatively civilised town, with fairly comfortable inns, and a considerable English society, including a bishop and his staff, and a sisterhood of English ladies to assist them. Moreover, it is the place which is put forward as an almost certain cure for all kinds of affections of the chest. I have before me a long list of cases who were sent there in an apparently hopeless state, and who are now enjoying fair health. I have no list of the cases who did not recover. It is probable that some unintentional exaggeration is used in speaking of Bloemfontein in the very exceptional terms in which it is spoken of at the Cape, and often in England; but there is no doubt that the climate is very fine, and that a great number of persons with serious diseases of the lungs have derived great benefit from a residence there.

Accurate meteorological observations would be of much value if they could be obtained, but no really complete records have been hitherto kept in Bloemfontein. I am told generally that the weather there in summer is excessively hot in the daytime, and too often also at night; and that the cold in winter is sometimes intense, the water frequently freezing in the bedrooms at night. It is said, however, that invalids do not suffer from these extremes in any very appreciable manner, as the air is remarkably dry, as would be expected from the situation of the place, cut off as it is by the Drakensberg range of mountains from the Indian Ocean on the east, and having a vast extent of dry country on the north and west.

Bloemfontein is a very good place from which to start on a wagon expedition through the Free State into the Transvaal. Wagons, oxen, and Kaffir boys are easily obtainable there, and there is no difficulty caused by difference in the grass on the high plateaux of the two States. For a description of the country I must again refer my readers to Mr. Trollope's book on South Africa. The months of March and April and September and October are the best months for such an expedition; but in ordinary seasons, and if care be taken to provide against extremes of heat and cold and heavy rain showers, there are few months during the year, except the height of summer and winter, when a wagon expedition might not be commenced with safety and comfort.

In making the necessary arrangements for an expedition of this kind, it would of course be wise to consult some experienced person on the spot as to the purchase and fitting up of the wagon, the selection of the oxen and horses, the engagement of the servants, and the line of country to be taken, with especial regard to the healthiest districts for man and beast, and, if the travellers be sportsmen, the best districts for large and small game.

I will however give here a few particulars to enable my readers to form some idea of the expense of such an expedition, and the sort of stores which are necessary to carry it out with comfort. The party should consist of two or three persons, all, I need hardly say, well acquainted with one another and able to put up with each other's occasional fits of moodiness or ill-humour. If they have a taste for sport, or natural history, so much the better for them, as the life is necessarily monotonous, and there are no grand ruins, museums, or picture galleries to be visited by way of change; and indeed towns or villages of any kind are very few and far between. In a full sized wagon there is good sleeping accommodation for three persons—two in the wagon and one in a hammock swung under it. It is not safe for an invalid to sleep on the ground, or without any covering overhead. A full sized wagon, called in the country a buck-wagon, is the most comfortable and most easily saleable when done with. It may be expected to cost, with all necessary fittings, from £120 to £150. It would require a span of fourteen or sixteen oxen to draw it; and these would cost from £5 to £10 each. A couple of riding horses, to vary the monotony of driving or walking, and for the purpose of riding after game, are a great convenience; but they are not necessaries, and they sometimes give trouble. They would cost from £15 to £20 each. The wagon and gear, and oxen and horses, if properly cared for and brought back in moderate condition, would all be saleable in Bloemfontein for nearly two-thirds of their original cost. Good second-hand wagons and gear can sometimes be bought at a considerably less price than the sum I have named; but great care must be taken to see that they are in thoroughly good condition for the journey, as a breakdown in the Veldt is a serious business. As to servants, a driver and a leader of the oxen are requisite; the latter looks after the oxen when feeding. A man who can cook is also a great addition, and he could also look after the horses. If no horses are taken, a cook can be dispensed with, but it entails considerable trouble on the different members of the party, if they have to cook their own food before eating it.

The men, or boys as all natives are called whatever age they may be, are all Kaffirs, and their wages would be about as follows:—The driver from £4 to £5 a month, and the others about £3 a month each, exclusive of food, which the travellers have to provide. If a man be a decent cook, he might expect rather higher wages.

Wagon travelling, with the traveller's own oxen, is very slow work, not averaging more than from ten to twelve miles a day; but as such an expedition as I have suggested is only made for the sake of health, the rate of travelling is not material; and when once a traveller has got into a district which is suitable to him, he is in no hurry to get out of it. A good supply of books should of course be taken; and some practical knowledge of skinning birds and preserving skins and insects is interesting and useful, as much has still to be learnt respecting the natural history of South Africa. All information respecting the requisite stores to be carried can be obtained on the spot; but tinned soups, meats, and vegetables, and preserved milk and butter can be purchased better and cheaper in England than in the colonies. Good English tea is also a great luxury. The travellers must make their own arrangements for the liquors which they severally affect, remembering of course that the less the bulk the easier the carriage. A good water filter is a most useful article; and the water which is drunk on the expedition should be both boiled and filtered beforehand, as it is often full of minute animalcules, which live and breed in the bodies of those who drink it. If all these precautions be taken, an expedition in an ox-wagon through the Orange Free State and the Transvaal may be confidently relied on as an enjoyable and health-giving way of spending three or four months.

I have said nothing of Natal or Griqualand West, where are the diamond-fields, as I cannot give any reliable information about either of them. From what I can hear, however, they are neither of them districts particularly fitted for invalids, especially those with delicate lungs.

Durban, the port of Natal, is a notoriously unhealthy place, and invalids who go there should remain as short a time as possible, and should go forward to Pieter Maritzburg for their head-quarters. A wagon expedition can well be organised and commenced there, and there is much less fatigue in getting there than to Bloemfontein, but I am told that the latter place is by far the better starting point of the two, as the traveller is in a high and healthy country from the very commencement of his expedition.

The Orange Free State and the Transvaal are the countries best situated for health resorts, and only require to be better known, and made more easily accessible, to become popular with both doctors and patients; though it remains to be seen how far the late disturbances in the Transvaal may make travelling or residing there impracticable for Englishmen for some years to come.

Since writing this chapter, I have read a book called "Eight Months in an Ox-wagon," by E. F. Sandeman, which I can cordially recommend to any of my readers who may contemplate such an expedition.

Mr. Sandeman was an invalid when he started, but the record of his doings after the first few weeks shows that he did not long remain so. The book gives a great deal of information, which would be most valuable to anyone organising such an expedition as I have suggested in this chapter.

In conclusion, I would strongly advise invalids who go to the Cape not to remain too long in the neighbourhood of Cape Town; and not to take up their abode for any length of time within 100 miles of the sea coast, nor less than 1,500 feet above the sea level. Unless these conditions be complied with I cannot think that persons with diseases of the chest or lungs would derive any great benefit from a residence in South Africa.

CHAPTER XI.

DAVOS.

Change of Treatment in Lung Complaints.—The "Germ Theory."—Necessity for competent Medical Evidence.—Earlier accounts of Place.—Situation and Surroundings.—Comparison with other Sanatoria. — Climate. — Cold and Dryness. — Toboggining. — Skating.—Weather.—Drainage.—Clothing.—Hotels.—Food and Wine.—Review of Earlier Accounts.—Effect of Climate on Invalids.—Whether Lasting or otherwise.—Causes of Favourable Results.—The Snow-melting in Spring.—Where to go when it Commences.—Route from England.—Conclusion.

No book which purports to give any account of the various sanatoria of the world, would now be complete without some notice of the newest, and, with many physicians, the most in fashion, of all the sanatoria. As medical science improves, and the laws which govern health become better known, it is inevitable that changes should from time to time be made in the different methods of treatment adopted by successive generations in respect of the same well known forms of disease. But surely no greater change has ever been made than that which substitutes, in the treatment of lung complaints, the intensely cold dry air of the high Alps in winter for the warm and genial atmosphere of places such as those I have attempted to describe, or like Madeira, or the well known sanatoria on the Riviera. If this substitution be based on sound principles, and be found to work well in practice, it is impossible to exaggerate the enormous benefits which will result to mankind from the discovery. I venture, however, to doubt whether there is not some slight element of fashion in the selection of a sanatorium for invalids, even amongst the ablest men in the medical profession; and whether this fashion does not vary from time to time, like fashions in dress, without, in all cases, a due consideration as to its suitability to the person who is advised to adopt it.

There seems to be at present a difference of opinion amongst the most eminent medical men as to the true causes and nature of the different kinds of lung disease; and there must consequently be a difference of opinion as to the various remedies to be applied in their treatment. One theory, which may be generally referred to as "the germ theory," has been most ably put before the world in an article by Dr. Clifford Allbutt, of Leeds, in the "Lancet" of 20th of October, 1877; and if that theory be correct, then nothing could be more conducive to the successful treatment of lung disease than a residence at such an altitude and under such conditions as prevent the generation of these germs and destroy them when generated. I believe, however, that this theory is very far from being accepted in all its fulness by the majority of the medical profession; and that many of the old ideas on the subject are still in vogue. What then will account for the extraordinary change in the treatment of the disease by some of those who hold

N

the old ideas? Is it that the new remedy has been found efficacious in practice? This is the question that I wish to bring before my readers; and I greatly doubt whether the answer can be as complete and satisfactory as the well wishers of Davos would desire.

It is true that the time during which the place has been in repute amongst English people has been very short, only indeed to any great extent since the winter of 1877-78. But during this period a great number of cases have been treated there; and it is of the highest importance to know with what result. I am aware that some of these cases, principally the patients of one eminent London physician, have been commented on by him in different numbers of the "Lancet" during past years. I have not had the advantage of seeing these comments, but I am informed that they were to a very great extent favourable to the remedial benefits to be derived from a winter residence at Davos. But what record is there of the patients of other physicians? and especially what record is there of those who have received no benefit from this treatment, but rather the contrary? That there were many such during the winters of 1878-79 and 1879-80, I can myself bear witness. What then were the causes which militated against the improvement of some invalids, whilst others who, to an ordinary observer, appeared in a very similar condition, seemed rapidly and steadily to grow better and stronger day by day? This is a question which requires a plain and decisive answer from some one whose opinion will carry conviction to the minds of those who are called upon to discriminate between the different modes of treatment which ought to be adopted in the different forms of lung complaints. If the claim which has been advanced in favour of the newer form of treatment can be established, then no time or trouble will ever have been better or more profitably expended than in establishing it; but if it prove illusive, then the sooner this can be done the better. My own very modest opinion, after a residence of two winters in Davos, is, that too much is claimed for it by its admirers, and far too little is conceded by its detractors. I believe, however, that no conclusion which will be really satisfactory to English people will be arrived at until some competent English medical man has spent one or more winters in Davos, and has taken particular records of a certain number of test cases amongst the invalids who are resident there, and has made known the results of his observations. The local doctors are so thoroughly assured of the enormous advantages of the place, that they would give every facility for this being done, so long of course as their own treatment of their patients be not interfered with. A knowledge of the German language would be a great advantage to any English doctor who would undertake this work, as he would thus have a far better opportunity of seeing a greater variety of cases, and of discussing the characteristics of each with the several doctors in Davos, who are all Germans or Switzers; but this is not essential, as there is one doctor there, Dr. Karl Ruedi, who attends nearly all the English invalids, and who understands and speaks English perfectly, and who is always able and most willing to give every assistance to any one who may be honestly endeavouring to satisfy himself of the advantages and disadvantages of the place.

If the report should be favourable, and I venture to think that in its essential features it would be so, then it cannot be too soon known to English people that there is a place at no great distance from home, whither they can transport themselves, with very little trouble and at no great expense, before the commencement of the winter, with the certainty of finding a fair amount of beautiful weather, and with a great probability that the invalid members of the family will derive material benefit from a residence there, and where the sound members of the family will probably improve in health and strength, instead of deteriorating in both, as is too often the case in warmer climates. If the report should be unfavourable, then many great invalids, who might otherwise be sent to places like Davos, on the chance of their getting some good from them, will be spared the misery of shivering through five long months in a climate, which in many winters is semi-arctic, and which may at the same time be doing them far more harm than good. Ignorant as I am of medical science, and unacquainted with the special characteristics of the cases of the different invalids around me at Davos, it was easy enough to see which of them seemed to derive real good from their residence there; which of them seemed to remain stationary; and which of them seemed to be going steadily downhill throughout the whole winter. I venture to think that hardly any of the persons in the last category would have been sent to Davos, if more real information had been previously obtained about the place. They passed, for the most part, a dull and monotonous existence, shivering from the cold both outside and inside of their hotels, and unable to take a sufficient amount of exercise to keep their blood in circulation; they suffered from increased difficulty of breathing, owing to the extreme rarity and diminished pressure of the atmosphere; and some of them got a catarrh over their chests, which seemed to hang about them during the whole winter. It may be that these cases would not have gained much good in any climate, but they would at any rate have passed a much happier existence amongst the olives, oranges, and flowers of more southern climes. On the other hand, I have met numbers of invalids, during my wanderings in search of health, who had been advised to winter in warm climates, and who had been nothing benefited thereby, but who would, I feel sure, have derived the greatest benefit from one or more winters at Davos.

It is the want of reliable information about the place which seems to me to be urgently needed by the medical profession at the present time; and principally information which can only be given by competent medical men who have resided there during the whole winter, and made minute observations upon a great variety of cases. It may be thought that this information might be equally well obtained by communicating with local medical men; but I venture to doubt whether this would be the case. There seems to be a considerable difference in the constitutions of English and German people; and many remedies which are adopted without question in Davos for the German invalids have been found to be of doubtful advantage for the English. Moreover, the local medical men have such unbounded faith in the efficacy of a residence at Davos in nearly every case of chest disease, that their opinion would hardly be accepted with that perfect confidence by English physicians, which it would be the main object of such information to supply. I cannot help feeling sure that there are many kinds of affections of the lungs, bronchial tubes, and throat, which for some reason or other are more likely to get harm than good from a winter residence at Davos; and, on the other hand, I am equally sure that there are a far larger number of cases, differing from the others perhaps in minute details, which would be likely to derive incalculable advantage from one or more winters spent there; and to whom a winter residence in warmer climates might be far less beneficial. The difficulty for the physician is to distinguish between these cases; and this can only be done after a more thorough knowledge of the place and its climatic influences has been obtained. I believe that no attempt has hitherto been made to supply anything like an exhaustive record of the English cases in Davos during any one of the winters they have resided there; and yet nothing would be easier to do, if only the proper steps were taken to obtain them.

I have read a pamphlet by Dr. Alfred Pope "On the Climate of Davos-am-Platz," which contains some useful information for invalids; but as he only resided at Davos from the 12th of November to the 10th of December 1879, which was confessedly the very worst period of a rather exceptionally bad winter, and as he does not seem to have examined any of the invalids with a view to reporting on their cases, this pamphlet does not supply to any appreciable extent the want of technical information about the place which seems to me to exist. I can only hope that some steps will be taken before long to supply this want. If no one medical man can spare the time to spend the whole winter abroad, it might surely be managed by three or four men who could take the work from one another.

Davos is by no means a bad place in which to take a holiday: and a great deal of good and useful work might be combined with a considerable amount of physical enjoyment. This I hope will be admitted after a perusal of what I shall here relate as my own experience of the place during the winters of 1878-79 and 1879-80. I think too, that a great deal of most useful evidence might be obtained if the authorities of the Brompton Hospital would send a certain number of their patients as test cases to Davos, as they did to Madeira in the winter of 1865. Proper accommodation and medical attendance would have to be provided for them, but the expense would not be very great, and the benefit to many of the sufferers, and to the world at large, might be incalculable.

There has been a good deal written about Davos lately, but mostly in a somewhat fugitive form, and not always easy to be laid hold of by the general public for reference when wanted. The first account in point of date was given in an excellent article by Dr. Clifford Allbutt in the "Lancet" of 27th October, 1877. The next, and perhaps the most important from the amount of information which it contains, is in the form of a small half-crown book, entitled "Davos Platz, a new Alpine resort for sick and sound in Summer and Winter," by an anonymous author, and published by Edward Stanford in 1878. Then followed a most admirable article in the "Fortnightly Review" for July, 1878, by Mr. J. A. Symonds, which has probably had more to do with bringing English people to Davos than all the other publications put together. There have been several letters, articles, and comments on cases in the "Lancet" during the years 1878, 1879 and 1880, which I have not had the advantage of reading, nor am I able to give references as to their respective dates of publication. Next came an article in the "Fortnightly Review" of November, 1879, by Dr. J. Burney Yeo, rather depreciatory of high Alpine health resorts in general, and of Davos in particular. And finally has appeared the pamphlet of Dr. Alfred Pope, to which I have before referred. I am not aware of any other publications on the subject in the English language, but much has been written about the place in German: and a list of these works is given in Mr. Symonds' article to which I have referred. I fear that it will be impossible for me to help travelling over some of the same ground which has been traversed by these several writers, but I will do it in as brief a manner as possible.

Davos-am-Platz is the principal village in the Valley of Davos in the Graubünden, and is situated at an elevation of about 5200 feet above the sea. The valley may be roughly described as about six miles in length by one mile in breadth, before it enters the narrow gorge of the Züge; and it lies nearly north and south. It is enclosed by wooded mountains of no great height, but sufficiently high to afford fair shelter to it from the cold north winds and from the relaxing south-west wind, which is there called the Föhn.

There is a lake at the head of the valley, whence issues a stream called the Landwasser, which runs through and waters the valley, and makes its exit through the Züge Gorge into the Albula, finally joining the Rhine at Thusis. The soil of the valley is gravelly and dry, except in the immediate vicinity of the stream.

I have never seen Davos in the summer time when the meadows are green, and the flowers, of which there is a great variety, gay and in countless numbers; but I can fancy that it would be a bright and cheerful looking valley. In the winter, when the whole surface is covered with snow, and when the surrounding mountains are white from the summit to the base, only contrasted, I can hardly say relieved, by the deep olive green of the fir woods which clothe the lower slopes of them, the aspect is somewhat sad; it may be even rather depressing to the wearied traveller on his first arrival on a cold grey evening in the late autumn, long after the last rays of the setting sun have bathed the whole valley in a warm golden light. Davos is, however, one of those few places in the world away from home in which the longer people remain the more they like it, until at last the great majority of the invalids with whom I was acquainted came to look on the place as a kind of second home. There were but few invalids of those who derived any benefit from the place during my first winter who did not return during the second: or who did not write sorrowful letters to their more fortunate friends regretting their inability to do so; and some of those who did not seem to gain much good from the place, and who left it before the winter was over, wrote to the friends whom they left behind them, lamenting their own shortsightedness in going away so soon, and announcing their desire and intention to return at the first opportunity. These all complained that till they got down into the Swiss valleys on their several routes, they had forgotten the experience of cold fogs and drizzling rain.

I may mention three cases of gentlemen who left the hotel in which I resided at Davos before the end of the first winter, one for San Remo, another for Cannes, and a third for Arcachon. The first said that he was confined to his hotel at San Remo for nearly a week immediately on his arrival by a strong mistral; and that the first time he went out afterwards he caught a bad inflammatory cold, which confined him to the house for a fortnight. The second said that he had only been once out of his hotel at Cannes for nearly three weeks on account of the biting winds, and then only for about twenty minutes, and he intended to go forward to Mentone, in the hope of a change for the better. And the third said that the whole country in the neighbourhood of Arcachon was under water and enveloped in a thick wet fog. They all complained that they had never felt the cold of Davos in anything like the same searching manner as in those three well-known sanatoria, though there was probably a difference of temperature in favour of the latter at that time of year of twenty or thirty degrees Fahrenheit at the very least. These men had all left Davos in disgust, and without feeling in the slightest degree bettered by it; and their evidence cannot therefore be said to be prejudiced in its favour.

It is of the greatest importance that invalids who go to Davos should have a fair reserve of physical vigour to enable them to resist the intense cold which they are almost sure to meet with during some part of the winter; but if they have that, they are far less likely to suffer from the miserable shivering sensations which accompany cold weather in lower regions, where the thermometer may register a much higher temperature. This is entirely to be attributed to the exceeding dryness of the air of Davos. Full records of the humidity of the atmosphere, and the ranges of the thermometer and barometer during the past three or four years, have been kept at the Hotel Belvédère in Davos, and can no doubt be obtained on application to Herr Coester the landlord, or to Dr. Ruedi, by any person really interested in these subjects.* It is sufficient for the purpose of this book to say that the only country in which I ever felt anything like the dryness of the air in good Davos weather was Upper Egypt.

The snow, when it falls, is like the finest sugar, and invalids may go out in it with impunity and without fear of a wetting, if only they will occasionally shake it off their outer garments. The skin gets harsh and cracks on the slightest provocation, though there is comparatively little cold wind; all leather articles are apt to crack in like manner; tobacco gets dry and dusty in a very short time, even when wrapped in lead paper; and during the winter of 1879-80, two sets of ivory billiard balls at our hotel split up, and the outer crust broke off in shavings, caused entirely by the exceeding dryness of the atmosphere. This kind of weather cannot be expected till after the first week in December, when the regular snowing-in has taken place; and it is afterwards liable to be frequently interrupted by periods of Föhn winds and thaw, which,

---

* I have lately been informed that these observations cannot be entirely relied on, as the instruments were placed under far too favourable conditions, through some mistake of the man who first set them up.

however, do not often last for more than three or four days at a time; these periods of Föhn are generally followed by the wind going suddenly and strongly round to the north, and bringing a snow storm which may last for a day or two; after which there generally comes another spell of beautiful dry weather for a longer or shorter period, according to the season.

I shall give here my own experience of the weather during the winters of 1878-79 and 1879-80, which were said to be typical, the former of a rather bad, and the latter of an extraordinarily good winter, in Davos; but first I must try to give some idea of a really perfect Davos day. There are probably from two to three feet of snow over the entire valley. The day has been preceded by a perfectly clear bright starlight or moonlight night, with not a breath of wind, and a minimum temperature of about zero Fahrenheit. After a night of this description, the early riser,—and every invalid who is in moderate health should rise at a reasonably early hour if he wishes to get all the good he can out of Davos,—will see a perfectly cloudless sky, with perhaps a faint remaining star, or the moon going down behind the mountains to the west. Soon he will see the highest peaks of the furthest mountains at the south end of the valley tipped with the earliest rays of the rising sun, which little by little will reach first one and then another of the lesser heights, losing in their course the soft rosy tints which they first had, and gathering whiteness and brightness as they at last extend over all the summits of the mountains which surround the valley. All this time the valley itself is in complete shade. In consequence of the configuration of the mountains, it is 10 o'clock before the sun first shows his head above the summit of the Jacobshorn during the months of December and January, and till then the thermometer rises but very few degrees above its minimum temperature of the night. Within two hours from that time it is not improbable that the solar thermometer with blackened bulb will register 120° to 130° Fahrenheit, whilst in the shade the temperature is still very considerably below the freezing point. Not a breath of wind is felt, not a branch or leaf stirs : but the air is clear and fresh off the snow, and there is not a particle of dust. In no country which I have visited have I ever seen anything more exquisitely beautiful in the way of climate than a perfect Davos day.

Long before noon nine-tenths of the invalids have turned out, the less vigorous of them to sit in the verandahs of the different hotels, with books or work; or to drive in sleighs along the smooth snowy roads to the different places of interest or beauty in the neighbourhood. These invalids are all well wrapped up in furs and rugs, but nearly all of them are compelled to wear sun hats, shades, and spectacles to ward off the burning rays of the sun. It is impossible to be too well wrapped up in taking a drive, as the motion through the air creates a current, and the least wind is keenly felt when the temperature is so low, and when the invalid is not in exercise. But on the best days all extra wraps have to be discarded when sitting in the verandahs; and the only protection required is from the sun. On these days the more vigorous invalids take exercise in proportion to their strength, and in accordance with their several tastes, from sunrise to sunset. Some walk on the level roads or up the pretty valleys which run at right angles out of the Davos valley, or up the steep paths which are kept in order by the Curverein amongst the fir woods of the Schatz Alp. Others skate on the ice rinks, or enjoy the still harder exercise of toboggining down the sloping roads and paths in the neighbourhood of their hotels.

Toboggining is an amusement which has only to be known to be appreciated. It is performed by one or two persons sitting on a small and low sleigh, which they have dragged up a steep incline, and shooting rapidly down the incline to the bottom. It requires practice, skill, and sometimes courage to guide these little sleighs, or toboggins as they are called, safely round sharp turns and down steep dips, especially when the male tobogginer has a fair female friend on the same toboggin behind him. Many are the upsets and rolls in the snow which everybody gets in turn, but I have seen but few accidents of any consequence; and strangely enough invalids who hate snow in their own country seem to like half burying themselves in it at Davos; and they seldom take any harm from it. It is perfectly wonderful how long many invalids will go on at this amusement without resting or appearing to feel much fatigue from it. It is hard work dragging a toboggin up the same incline six or eight times in succession, but the excitement of the descent seems to make up for all the toil. When the meadows are frozen sufficiently hard on the surface to admit of toboggining over them, the pleasure is enhanced tenfold; and I have often been out for a couple of hours on fine bright moonlight nights, and have even got up at 6 o'clock in the morning, to enjoy this amusement in perfection. Numbers of invalids, both male and female, do the same thing, and seem all the better for it. Meadow toboggining however requires a combination of thaw and frost, which rarely occurs before March, and which sometimes does not occur at all.

Toboggining is a national amusement of the inhabitants of the district, who have a pattern of their own in the shape and size of their toboggins. They differ from those in use in Canada, and are much better fitted for the requirements of the place. Toboggins of a larger size are used by the inhabitants for bringing down the hay and fodder, which have been made and stacked on the mountains during the summer, into the valley for the consumption of the cattle which are all stabled during the winter, and for loading fuel and any other articles which may be wanted, and which have to be brought down the steep sides of the mountains.

There are now two very fair ice rinks close to the village; and each year a greater amount of care and attention is bestowed on them to keep them in good condition: they are generally ready about the middle or end of November; and skating continues till about the first week in March, not however by any means without interruption, as there are frequent heavy snow-falls which often over-tax the energies of the sweepers who are employed to keep the rinks in order. In some exceptional years there are a few days of splendid skating to be got on the lake; but it too often happens that during the short time that is requisite for freezing the surface into a sufficient thickness for skating, there comes a fall of snow which spoils it for the whole of the rest of the winter. I think however that a great deal more might be done than is done towards clearing and keeping cleared a sufficient space of ice on the lake to admit of excellent skating, if proper measures were taken beforehand.

We had a few days' skating on the lake during the month of December, 1879: the ice was black, and in some places as smooth as glass; but the weather was unfortunately very bad. There was a strong and bitter north wind blowing the whole time, with thick clouds on the mountains, and a light driving snow in the air. The thermometer rose but little above zero Fahrenheit during each day, and fell to fourteen or fifteen degrees below zero during three successive nights. We were of course all well wrapped up in thick coats or sealskin jackets, and our hands encased in fur gloves; but I never felt the cold in Davos in anything like the same degree as I did during those days on the lake. Those of the ladies who had veils had to take them off, as they soon became one sheet of ice, and icicles hung from the moustachios and beards of those of the skaters who had such appendages. Our handkerchiefs froze so hard in our pockets that they had to be constantly thawed at a fire which we made for the purpose; and on one occasion my handkerchief absolutely broke into two pieces like a piece of thin shaving. It would be considered by most people in England to be almost madness in any of the invalids to be found skating on the lake, which is in a particularly exposed situation, during such weather: but many of the stronger ones were skating there for three or four hours on each day; and I heard of no one being any the worse for it, except one or two men who had a slight touch of frostbite. I had had a cold hanging about me for some time, and I completely lost it on the lake during those days.

I find that I have made a rather rapid transition from a description of the best days to that of the very worst days in Davos; but there are a great many days during every winter which do not come within either category. The weather during the winter may be roughly divided into four classes, perfect days, fine, indifferent, and bad. A perfect Davos day has been already described, and I shall now give some idea of what I include in the other three classes. Fine days are those which would be called very fine in most other parts of the world, but which are not considered perfect in Davos by reason of there being a little wind stirring, or some light clouds in the skies. These are often very pleasant days, and the beauty of the mountain scenery is enhanced by the light fleecy clouds which hover amongst the highest peaks; but complaints are sure to be heard that the sun was obscured for five minutes whilst the complainant was sitting in the verandah; or that the light wind, if from the north, was too cold, or, if from the south, was too relaxing, to be endured by anyone. We all seemed to forget that sitting in the open air without any great coat or rug, between the 1st of November and the 31st of March, is an unusual occurrence in Europe, especially at an elevation of over 5,000 feet above the sea. Indifferent days are those when there is no bright sunshine, and clouds are upon the face of the earth: there is probably a light snow falling; but there is no wind, and the thermometer registers a low temperature. Invalids in fair health may then take exercise without fear of harm, and generally with positive benefit. The air may be dry in spite of the clouds and snow; and the absence of wind prevents any feeling of chill to the walker, or any suspicion of catching a cold; but it is not weather for sitting in the verandahs of the hotels. In this class may also be comprised fine days when there is a light Föhn wind blowing, and a consequent melting of the snow and dampness in the air. To some few invalids this change from the extreme dryness of the atmosphere to comparative dampness and softness is grateful and beneficial, especially in cases of bronchial irritation; but to far the larger number the change is prejudicial. Even when the Föhn blows lightly it produces a certain amount of lassitude and enervation, and frequently rheumatism and neuralgia amongst both weak and strong people. If it continues and gets stronger it is also apt to produce disorders of the liver and digestive organs, and general suicidal tendencies; but these are all dispelled on the first return of the wind to the north, which seems to act like magic even on those confined to the house. Bad days are those on which there is either a strong north wind, accompanied or not with snow, or a strong Föhn wind blowing in the valley. On these days all invalids should remain indoors. There is indeed no temptation for them to do otherwise; and this is one of the advantages of Davos.

In many of the better known Sanatoria how delusive is sometimes the appearance of the day from the south windows of the hotels. There may be a clear blue sky and a bright sun and a warm air on the sheltered side of the house; whilst one of the many bitter winds of the north shore of the Mediterranean may be blowing, or preparing to blow, a gale in all the exposed parts of the district. Woe to the invalid who, tempted by the fair exterior, rashly ventures out for a walk or a drive on such a day without his fur-coat and wraps. It is well for him if he be only confined to the house for the next fortnight with a bad cold in his head. There is no such inducement at Davos: the appearances of the weather are very seldom treacherous.

If there be a strong north wind blowing, the sky is black, and clouds cover the mountains all around the valley: the snow is whirled from the ground in eddies round each corner of the hotel, and the trees shake off their snowy burdens in perpetual avalanches. There are the same appearances in the sky and on the mountains in the case of a strong Föhn; and probably in addition, long wreaths of mist enveloping the valley; the snow is thawing rapidly, and there is a steady dripping sound from every pipe and spout within hearing. These sights and sounds are unmistakeable by any one, and, when they appear, the invalid settles himself down quietly to his indoor occupations. If the strong Föhn continues, the snow begins to melt off the sides of the hills, and the roads become, first a mass of slush, and then a running stream, into which every house and *châlet* pours its contributory streamlet. This is fortunately but a rare experience during the winter: it may be seen in November, but probably not again till the end of February at the earliest. It is then the beginning of the end: there will be more snow and frost and many more brilliant days, but the great snow-melting will be considered to have begun; and nothing will be heard but schemes of travel for the inevitable time when all visitors are advised that they must leave Davos.

The Föhn wind, which answers to the Italian Sirocco, is a hot and exceedingly dry wind. There is a conflict of opinion as to whence it comes; but the common idea is, that it comes from the centre of

Africa, and gets its heat and dryness from blowing over the vast tracts of desert. Most travellers have felt the evil effects of this wind, under whatever name it goes, in the south of Europe. It is especially bad in Naples, Rome, and Florence, and along the Riviera, where it is so much dreaded by the invalids. Dry though the wind is, it almost always leaves behind it signs of dampness on the pavements in the streets, which is accounted for on the supposition that it takes up the moisture from the Mediterranean in its passage across the sea and deposits it on the land in its northward course. This wind is often felt strongly in Switzerland, and is the cause of most of the fires which have so frequently destroyed Swiss villages, in spite of every precaution which is taken to prevent fires being lighted during the prevalence of it. It is impossible to shut the Föhn wind out entirely by any barriers; but it is obvious that its force may be broken by sufficiently high mountains, and that some of its most pernicious effects may be mitigated by passing over a large region of snow. This is one of the advantages which is claimed for the valley of Davos. The Föhn is felt and sometimes felt strongly there; and I have not hesitated to state the bad effects which it produces; but I know of no part of Switzerland which is more free from it, nor where its effects would be likely to be less pernicious. The usual period of its continuance is three days: it sometimes extends to the fourth day, when the wind generally makes a sudden shift to the north, and brings on a snowstorm and subsequent change to fine bright weather. The longer and stronger the Föhn wind blows the worse of course are the ill effects to the invalid: and to these must be added, when a thaw sets in, the great additional chances of colds and catarrhs and their attendant evils; and the greatest precautions must be taken to avoid them.

I shall now state the number of days of each kind of weather which I experienced during the two winters which I spent at Davos: and I am fortunately able to give, on perfectly reliable authority, similar details of the winter of 1877—78.

If an average be taken from the records of these three years I believe that a fair idea may be obtained of the probable weather of an ordinary winter at Davos. The two winters of 1877—78 and 1878—79 were considered to be rather exceptionally unfavourable at Davos, but the winter of 1879—80 was a very exceptionally good one, no similar winter having been experienced there since the year 1873—74. The following table will show at a glance the weather in the three winters to which I refer.

1877—78.

Date.	Perfect.	Fine.	Indifferent.	Bad.
November	3	9	12	6
December.	5	7	13	6
January	7	9	10	5
February .	13	3	12	0
March .	3	8	11	9
Total . .	31	36	58	26

1878—79.

Date.	Perfect.	Fine.	Indifferent.	Bad.
November	9	8	6	7
December	1	12	15	3
January	12	10	8	1
February	5	4	12	7
March	9	11	7	4
Total	36	45	48	22

1879—80.

Date.	Perfect.	Fine.	Indifferent.	Bad.
November	6	4	12	8
December	18	2	5	6
January	16	7	6	2
February	11	8	9	1
March	16	9	2	4
Total	67	30	34	21

In the winter of 1877-78, there were four days only on which rain fell, namely, the 10th, 11th, and 28th of November, and the 29th of March; and on the 4th and 5th of December there was fog. In the winter of 1878-79, there were also four days on which rain fell, namely, on the 27th and 28th of November, and a few drops only on the 10th and 11th of February. I have no record of fog in that year; but it often happens that a fog hangs over the Landwasser even in the finest weather till the sun gets full power over the valley. This fog, however, very rarely extends for more than 100 yards on each side of the stream; and no one need go near it, except he be going to the ice rinks, which no invalid should do until the fog be quite cleared off. In the winter of 1879-80, there were only two days on which rain fell, namely, on the 4th of March, when it rained steadily nearly all day, and during the night of the 5th of March, when it rained heavily enough to spoil the surface of the snow on the meadows for toboggining for the rest of the season. It must be understood, in looking over the above weather table, that what I call perfect days are such as I really believe I never met with in any other place than Davos, combining as they do the extremes of brightness, freshness, sunheat, and freedom from dust and other drawbacks of hot climates. What I call fine days, would be considered very fine days in most other parts of the world; indifferent days would be considered fine in England and in other less favoured countries; and even the bad days are not half so bad as bad days at home. We never had anything to compare with a rainy winter day, or bitter spring north-easterly day, in England.

The figures for 1877-78 are from the same record on which Mr. Symonds based his remarks on the weather in his article in the "Fortnightly Review;" and yet he adds, "There was not a single day in the whole winter on which I was debarred from taking a moderate amount of exercise; and on a large majority of days I spent from 9 A.M. to 5 P.M. in the open air, partly walking and partly sitting when I was not driving, often adding a walk at night before bedtime." Mr. Symonds was at that time in a very delicate state of health, and exceedingly susceptible to colds. I was a comparatively sound man, and I had left England more as a matter of precaution and habit than for any other reason, so that my experience in this respect is not of nearly so much value. I may say, however, that so far as I remember, there was no day during the two winters which I spent at Davos on which I should have been the least afraid to take a walk for a couple of hours at some time or other. Indeed, the days in the early part of December, 1879, on which I and many very much more delicate people than myself, skated on the lake without taking harm, as I have before mentioned, were amongst the worst days I ever remember to have seen at Davos. There were many days during each winter when I did not go out of the house, but only because the outlook was not tempting, and we had other ways of taking exercise and amusing ourselves indoors. I think, however, that prudent invalids should count on remaining indoors on most of the days which I have marked in the table as bad.

There is no doubt that the last of the three winters to which I have referred, was by far the most enjoyable; but it seems doubtful whether it was the most beneficial in its sanitary results. The great desideratum at Davos is dry snow and plenty of it; and in this respect the first winter contrasted favourably with the last. Mr. Symonds says in his article that his own experience of that winter led him to expect two snowy days to three fine ones, and yet he adds " snowfall is, however, no interruption to exercise, and I never found that my health suffered from bad weather; on the contrary, I had the exhilarating consciousness that I could bear it, harden myself against it, and advance steadily under conditions which in England would have been hopeless." The two last winters, on the contrary, were remarkable for the small quantity of snow which fell during them, as compared with most winters at Davos. This was the more singular, as throughout the whole of the rest of Switzerland, the snowfall was far in excess of the average. We read in the papers that all the high passes were snowed up and the mails stopped many times during the winter, whilst at these very times we were wishing for snow to cover the hillsides, and to make the roads more practicable for the sleighs. We often got the clouds on the mountains, which told of bad weather in the lower regions, but we had nothing like the average of snowy days of which Mr. Symonds speaks in the preceding winter. The result was, I think, more dampness in the atmosphere during the second winter, and certainly less successful results in some cases than in the preceding winter. The last of the three winters was for the most part beautiful and very dry, but I cannot say that I noticed that great improvement in the more serious cases which I should have expected from the weather which we enjoyed. I believe, too, that this was the experience of the local doctors. Various reasons were given for this result; one was that the invalids were tempted by the brilliant weather to do imprudent acts, which they would have had no temptation to do in less brilliant weather. And another was, that some of them positively suffered from the extraordinary dryness of the atmosphere, and from excess of sunshine and heat. This will seem a ridiculous statement to those who have never been at Davos; but it is nevertheless perfectly true that, in spite of the extreme lowness of the temperature in the shade during a considerable part of that winter, I heard many more complaints of the heat than of the cold. I, myself, delight in the sun; I have seen and loved him in many parts of the world, but I sometimes positively feared him when sitting in the verandah in front of our hotel during some afternoons in the months of February and March, 1880. Many persons who tried to take their usual exercise on these days returned home quite exhausted, and suffered afterwards in consequence. Others got heated, and then cooled down in the shade without changing their clothing, and soon got chills and colds; and others again, sat basking in the sun's rays till he went down behind the Western Mountains, and then went out for a walk at the moment of the greatest change of temperature without sufficient wraps, and naturally suffered in consequence. There is no dampness on these days at sunset, but there is often a fall of from 30° to 40° in the temperature in the course of an hour. Invalids should go indoors half an hour before sunset; but they are told in Davos that they may go out for a walk an hour afterwards; and they are advised to take a walk in the fine evenings after dinner, if well wrapped up. I confess to being sceptical about the advisability of making a habit of being out at night, except in cases of sleeplessness, which seem to be benefited by it. On the shortest days in winter, there are five good hours of sunshine, from 10 A.M. to 3 P.M.; and, if they be properly utilised, I think they should be sufficient for most invalids.

Many of us took our luncheons to the skating rink on the fine days throughout the winter, and ate them leisurely in the open air, sitting on benches amid the ice and snow, without thought of great coats, shawls, or any extra wraps whatever. During the whole of the best of these days the temperature in the shade was considerably under the freezing point, and, in spite of the hot sun, the snow showed no signs of thawing whatever.

It is a singular fact, that no matter what the temperature outside of the house may be at night, it seems to have but little effect on the temperature in the bedrooms, even though the windows of the rooms be open. I had a thermometer in my bedroom, and another hung on a nail just outside my window, so that I had every opportunity of satisfying myself on this point. All the rooms have double windows, the inner one being made so that the top part, about a foot and a half in depth, may be allowed to fall back on a hinge. If the outer window be left a little open and the top of the inner window be let down, the fresh air gets into the room without a possibility of draught to the person in it. It is strongly advised by the local doctors, that this should be done at night by all invalids, unless there be a dampness in the air. I always did so, and found it pleasant and healthful, though I cannot sleep with my window open in England without getting cold and sore throat. I had a fire lighted about 3 o'clock each afternoon, which was allowed to go out before 6, having thoroughly heated my stove, which remained heated for many hours afterwards. On the coldest nights the thermometer in my room when I went to bed would register from 50° to 53° Fahr., and it would be within two degrees of the same height when I got up in the morning. On these nights my thermometer outside the window often registered from 2° to 3° above zero Fahr. to 15° and 16° below zero; and on two occasions it went down to 20° below zero without making the slightest appreciable difference in the range of the thermometer inside the room. This was the experience of all the persons in the hotel, though of course the thermometers in their respective rooms varied in proportion to the amount of fires which they kept up, and in accordance with the aspect of the rooms, whether they faced to the north or south. We could only account for this from the fact that the air was perfectly still outside, and there was no cold wind blowing into the rooms, and the hotels were well warmed throughout, and the stoves and inside walls had never time to get thoroughly cold before being reheated in the morning. Whatever, however, may be the reason, it is a fact which is most important to invalids, that in the worst weather they will have no difficulty in keeping up an even temperature in their rooms, with a very moderate expenditure of fuel, assuming that they have rooms with a southern aspect, which all invalids ought to have. Most people in Davos find that a temperature of from 50° to 55° Fahr. is as much as they can bear in their bedrooms with any comfort. A temperature of 60° Fahr. at night feels far too hot and stuffy. I am aware that some of these statements, if accepted at all, will be accepted with reserve by most of my readers who have never been at Davos; and the life there does sometimes seem incredible even to those who are experiencing it. I can only refer persons who may suspect exaggeration in these statements to any friend who may have spent a winter in Davos, and especially to those who were there in the winter of 1879-80.

Davos is now a large village, and it is increasing rapidly each year in size and importance. The houses are large and airy, and the streets are fairly clean, except during a thaw. There is not however an entire absence of unpleasant odours from the open drains at certain times, nor a perfect freedom from refuse matter thrown out by some of the inhabitants into the public streets in front of their houses. The drains are dependent for their flushing out on running streams from the hill-side at the back of the village; and in the depth of winter these streams get almost completely blocked with ice, and the slender trickling stream which may possibly remain open in the midst of the ice is wholly insufficient to carry off the drainage. I believe that hitherto Davos has been a very healthy village, and I never heard of any of the numerous diseases which attend on bad drainage being epidemic in the place. I think however that, as the village increases in size and in the number of its inhabitants, the authorities ought to bestir themselves more than they do to put down the nuisances which will no doubt otherwise increase year by year. It is as much in the interest of the place as of the general public that I would remind the owners of property in Davos that one scourge of typhoid fever would in all probability ruin the place for ever, especially amongst English people.*

There are good shops in the principal street, in which nearly everything that may be wanted by an English visitor can be procured, from a suit of clothes to "Wills's Bristol Bird's Eye" tobacco. It is however advisable for smokers to take out their own tobacco and cigars, as the Davos tobacco is too often merely dry dust, and most of the Havana cigars come from Germany.

Cigars and tobacco can be imported into Switzerland on payment of a very small duty; and the latter should be securely packed in tin canisters or lead paper to prevent it becoming too dry.

English visitors will also probably prefer to get their clothing in England, and it may be well to mention the kind of clothing which seems to me most useful. Good warm and new woollen underclothing of all descriptions is very requisite : the upperclothing may be such as is usually worn in winter in England. Plenty of warm woollen stockings and socks, and one

* Since writing the above, I hear that this subject of drainage has become a very important one in Davos.

or two pairs of thick laced boots which come well up over the ankles to keep out the snow are essential, as well as warm gaiters for both men and women. It is of the utmost importance to invalids to be thoroughly protected against every kind of chill from the top of the head to the sole of the foot. Warm and snow-proof overshoes, invented, I believe, in America, and there called "Arctics," can be purchased at Davos, and are commonly worn by ladies who cannot wear boots thick enough to keep out the snow.

It is certain that invalids will feel the cold more or less, especially at the first commencement of the winter, when there is always a considerable amount of dampness in the air. It is therefore of great importance that they should get to Davos early enough in the year to become to some extent acclimatised to the place before the autumn weather begins to break up, which may be looked for towards the end of October. I think that no very delicate person should go later than the first or second week in October; and it would be much better to go earlier. The late summer and early autumn are often very fine and pleasant in the high Alps, and a fair stock of health and strength may be laid in to contend against the bad weather which may be expected in November, which is the most trying and disagreeable of the winter months at Davos. Moreover, by going there early, invalids would have time, in case they found the place unsuitable, to make their arrangements for going elsewhere before the real winter sets in.

There are several good hotels at Davos, suitable to the wants and pecuniary capacities of the visitors of the different nationalities who go there. The principal ones are the "Kurhaus," "Rhätia," "Strela," "Post," "Schweizerhof," "Buol," "Belvédère," and the "Neu Belvédère," now called the "Angleterre." Of these the first five are mostly frequented by Germans and persons other than Englishmen, though there are many English people who go to the Kurhaus and Rhätia, in preference to the purely English hotels, for the sake of getting a little foreign society. I believe that both of these hotels are comfortable; and they have the reputation of being well managed. The reception rooms at the Kurhaus are good and well fitted up, and many of the bedrooms, though rather small, look thoroughly clean and comfortable: it has, moreover, the advantage of having two or three large houses in its immediate neighbourhood which belong to it, and where families can get suites of apartments, which would be more likely to be quiet and retired than in the hotel itself. I had friends at both the Kurhaus and Rhätia, and they all expressed themselves as quite satisfied with their quarters. I think the great disadvantage of both of these hotels is the system of heating, which is done by steam. I always felt a sense of oppression when I went into the sitting-rooms, and a general want of fresh air throughout the hotels.

The three last named hotels, Buol, Belvédère, and Angleterre, are the three essentially English hotels; the two latter belong to the same proprietor, though they are under different management. These three hotels are situated a little distance out of the village, and they have their fronts well exposed to the sun. They appear to be solidly built, with double windows for the winter months. They are warmed throughout with porcelain stoves, which can be regulated at discretion, and which seem to me admirably adapted for their requirements. I never suffered in our hotel from overheating or closeness; and yet a comparatively small amount of fuel would soon make up a capital fire, and heat the rooms to any temperature that might be required. The food and general arrangements at these three hotels were very similar; and, if I give the preference to the one in which I spent two pleasant winters, it must not be taken as in any way a disparagement of the others.

I was at the Hotel Buol; and in all essential particulars I found it thoroughly comfortable. It is a large hotel capable of accommodating from seventy to eighty persons, and the situation is, to my mind, by far the best in Davos. The sanitary arrangements are good, and the hotel is plentifully supplied with excellent water, which is brought from a spring in the hill-side behind the house. The dining-room is large, airy, and comfortable, and the general sitting and smoking-rooms are sufficiently good. The bed-rooms are for the most part of a fair size and well ventilated, and the house is well furnished throughout. The proprietor, Herr Buol, belongs to one of the oldest and most respected families of the district, and he quite acts up to his position and reputation. He has a competent staff of officials to manage the different departments of the house, some of whom speak and understand English perfectly. It says well for the management that, during the two winters I was there, we never had any complaint to make of any serious nature; nor did we ever make any reasonable requirement that was not courteously attended to, and as far as possible complied with. We had fortunately a very pleasant society amongst ourselves during both winters, and I can unhesitatingly recommend the hotel to either single persons or families as a comfortable winter home.

I cannot leave this subject without referring to the question of the food which is provided at the hotels, and which I understand has been very unfavourably commented on in some quarters, and has led many English physicians to hesitate before sending their patients to Davos. There were, no doubt, complaints from time to time during the two winters I was at Davos, that the dishes were often too rich for English tastes, and especially for invalids; but it must be remembered that it takes some time to teach foreign cooks that English people like their meat as plainly cooked as possible, and that they want nothing more than good beef, mutton and poultry. I hear that this lesson has been thoroughly learnt in the English hotels at Davos now, and that the food of all kinds is now of as good quality as it always was of amply sufficient quantity. I have had some experience of pension arrangements in France and Germany and Switzerland, and I think the food at Davos would compare very favourably with that of most other places of the same description. The pension prices were very reasonable, ranging from 7 francs to 12 francs a day for each person, according to the size and position of his room. I may add one piece of evidence that the food, even in past years, cannot possibly have been as bad as some persons have tried to make out. There were but very few people in our hotel during the two winters I spent there who did not increase steadily in weight, month after month, during the whole winter. Nearly all those who profited by their winter residence there gained from four to fourteen pounds during the five months: one young man at our hotel gained over two stone during my first winter; and a young English lady at another hotel gained twenty pounds during the second winter. I gained four pounds during my first winter, and about two pounds during my second. In every preceding winter that I spent out of England I invariably lost weight, and sometimes to a very considerable extent.

There is fair wine to be got at reasonable prices; and in this respect the hotels in Davos compare most favourably with all the hotels in the Riviera. It is probable that the neighbourhood of Cette may account for the badness of the wine in the Riviera; whilst Davos has the advantage of being in the immediate neighbourhood of the district where are made the celebrated Veltliner wines, and it is well supplied with them.

They are all sound in quality and very palatable, when persons have once got over the somewhat peculiar rough taste which is very noticeable at first. When these wines have been kept for a year or two in bottle, they deposit a thick crust, and have the taste and colour of very old port wine. These wines are highly astringent, and are considered to be very wholesome to persons who are debilitated by lung disease, especially if suffering from diarrhœa: and they form one of the few remedies which are prescribed for the patient on his arrival at Davos.

It used to astonish persons who came as visitors to Davos during the winter to see the invalids assembled at table d'hôte in the different hotels doing full justice to the food and wine before them. As all the residents have been together for some time and have become more or less intimate with one another, there is a far freer flow of conversation and hilarity than at ordinary public hotel dinners; and there were very few of the assembled invalids who bore any of the external signs of illness. They look for the most part sunburnt and robust, and the people who looked ill were the visitors, especially if they came direct from work in England: their faces seemed unnaturally pale and sickly in comparison with those of some of even the less vigorous invalids, and they could be hardly made to believe that they were in a colony of consumptives. It was, no doubt, too often the case that these appearances were very deceptive, and that the healthful hue on many of the faces was but skin deep. Still, I think it is an advantage, in a place where so many invalids are brought together, that they should appear to one another moderately well and robust, in place of presenting the white and deathlike appearance which they so often do in many other regular sanatoria.

I shall now touch very briefly on the previous accounts of Davos to which I have above referred; and I shall then state as shortly as I can the kind of cases which seemed to me to be fitted, and those which seemed to me to be unfitted, for a winter residence there. I shall do this with considerable hesitation, knowing my own incompetency for the work; but I shall do it in the hope that something I may let fall may act as a hint to those who have the knowledge and ability to distinguish between what is sound and what is unsound in my conclusions.

I have not a copy of the "Lancet" by me containing Dr. Allbutt's article on Davos; but, so far as I recollect it, he speaks more from the theoretical point of view than from any practical knowledge of the place as a winter health resort, he having only been there for a few days in the summer. He collected all the therapeutic facts which he could from the local doctors; but, though those who know these doctors may be fully satisfied of their skill, and the accuracy of their statements, it is inevitable that their evidence will be received by the general public with a certain amount of suspicion that they may be prejudiced in favour of their own sanatorium. Moreover, at the time when Dr. Allbutt wrote his article, the number of English patients at Davos had been so limited that it was impossible to arrive at any definite conclusions as to its suitability for English constitutions. It would be interesting to know whether Dr. Allbutt has seen any reason to modify the very favourable views which he then expressed, after a consideration of the cases of those of his own patients whom he may have since advised to winter there.

The book called "Davos Platz, a new Alpine resort for sick and sound in summer and winter," was written by a lady who had been residing at Davos for many years, and who has derived the greatest benefit from her residence there. It was written, I believe, after an unusually fine winter, and in the first exuberance of spirits consequent on a greatly improved state of health; and it may thus perhaps appear to a prosaic patient during an ordinary Davos winter, to be a little idealised. It contains, however, a large amount of most valuable information about the place and its resources; and many most useful hints as to the requirements of English people who may be meditating a visit there either in winter or summer.

The article by Mr. Symonds, in the "Fortnightly Review," on the other hand, though written in similar exuberance of spirits, seems to me to contain nothing but simple facts, which may be verified by any invalid who spends the winter in Davos. The winter of which he writes was by no means an exceptionally fine one; indeed, as he says, his own experience led him to expect two snowy days to three fine ones; and he describes in no rosy colours the necessarily somewhat monotonous life of the place. In spite, however, of these drawbacks, he steadily improved in health during the whole winter; and, though he seemed to fall back somewhat during the succeeding winter, consequent in a great measure on a bad cold which he caught in Italy during the autumn, and which he never seemed thoroughly to shake off, he improved again greatly during the following winters; and he is now able to do a fair amount of physical work, and a large amount of mental work, nearly every day. He himself says that he was never in any place where he could do a greater amount of hard work at so little expenditure of vital force as in Davos. His descriptions, too, of the place and scenery in fine wintry weather, though poetical, are strictly accurate; and they bear out what I have attempted to say in a prosaic manner in the earlier pages of this chapter. At the same time, I know that Mr. Symonds agrees with me in thinking that more accurate statistics should be obtained of the cases which have been already treated in Davos, before sending patients there in the somewhat promiscuous manner which now seems to be coming into fashion in England; and he now fears that the place is getting too much populated to possess all the advantages as a health resort which he considered it to have at the time he wrote his article.

It was a somewhat refreshing change, after hearing so much in favour of Davos, to read the rather depreciatory article in the "Fortnightly Review," by Dr. Burney Yeo; but it is a question whether, in his desire to state the other side of the case, he has not unwittingly rather overstated it. I am not presuming to criticise his medical opinion; but I think he has listened rather too readily to any unfavourable statements about Davos. For instance, he quotes without comment the statement of a "foreign gentleman" who had wintered in Davos four or five years previously, who told him "that during the winter he spent there, he frequently saw two and three interments a day; his wife who went there to be with her husband maintains to this day that she acquired a severe gastric catarrh there, which made her life a burden for years, and from which she has hardly yet recovered completely." Now I have before me a certified copy of the register of burials at Davos of the visitors of all the different nationalities for the period commencing on the 1st of January, 1875, and extending to the 31st of December, 1879, and the number amounts to 118. These are distributed as follows: seventeen in 1875, twenty-eight in 1876, twenty-six in 1877, thirty-two in 1878, and fifteen in 1879. Of these only five were English, one in 1877, three in 1878, and one in 1879. It will be observed that the statistics for 1879 comprise only about half of the winter season of 1879-80; but I have no reason for thinking that the latter half of that season was more prolific in burials than the former. I myself only saw two funerals during the whole of my two winters at Davos. Considering that during each of those two winters there must have been between 500 and 600 visitors in Davos, many of them going there in a very precarious state of health, it seems to me that the number of burials is extraordinarily small. The winter of 1880-81 was an exceptionally fatal one in Davos; I have not been able to get statistics of the number of deaths during that winter, but I have heard on reliable authority, that they were much above the average of the five preceding years, especially amongst the English visitors. I believe that the deaths in the winter of 1881-82 were not above the average.

It is true that the figures in the above register do not include the inhabitants of the place; but I may safely say that if they were included, they would have made a considerable reduction in the percentage of burials. The inhabitants seem a wonderfully healthy race, and the children look especially robust and cheerful. I am assured that pulmonary consumption is unknown as a disease in those high valleys.

I think it is not improbable that, if the "foreign gentleman" made a slight exaggeration in his statement about the burials, his wife may easily have done the same about her "gastric catarrh," though that is not so easy to prove. I venture, however, to say that, if there be one complaint which above all others seems to be benefited by a residence at Davos, it is a disorder of the stomach and digestive organs. I can speak for myself that I have never been better in this respect than at Davos; and

I have reason to complain sometimes of these organs in other places. In spite, however, of what seems to me this slight defect in Dr. Yeo's article, there is very much in the caution which he gives against running into extremes in favour of high Alpine winter resorts without due consideration, which commends itself to any unprejudiced person who may have tried them. Dr. Yeo gives some interesting advice about the winter climate of Egypt, which he compares in many ways to Davos ; but I think he can never have been at either Helouan les Bains, or Ismailia, or he would hardly have recommended them as " agreeable and advantageous " places for invalids in which to spend the winter.

Dr. Pope, in his pamphlet, states some useful facts about Davos; and he gives some opinions as to its suitability or otherwise in different cases of disease, and some practical hints for the guidance of invalids going there. I cannot presume to criticise his medical opinion on the subject ; but I should have ventured to doubt whether there are not other cases, besides those to which he refers in his pamphlet, which would be more likely to be benefited by a residence in a milder climate than Davos. He seems to think that, if the heart, brain, and spinal cord, are in a quiescent and normal state, nearly every other kind of disease would be suitable for treatment at Davos, and likely to be materially benefited by a residence there. If Dr. Pope had remained longer than he did at Davos, and had watched other cases during the whole winter, I am doubtful whether he would not have had cause to modify this opinion to some extent. There seems to be no doubt that persons suffering from affections of the heart, brain, or nervous system should not be sent to Davos; and I believe it is not considered a suitable place for persons with any serious derangement of the liver. It has been said that it is only organic, as opposed to functional, diseases of these organs which are unfitted for Davos, but this seems open to doubt. It is said that chronic affections are the complaints which are most successfully dealt with; but it is doubtful whether chronic bronchitis, especially in persons of mature or advanced age, is not increased, rather than diminished by the dry air of Davos. The place has been found beneficial in cases of bronchitis and asthma in children; but I heard nothing but complaints from grown up persons of irritation in their throats and bronchial tubes, which caused them more discomfort from frequent fits of coughing both by day and night than they ever experienced even in England.

There seems to be something peculiarly irritating to the throat in the air of Davos. Many patients who were advised that they were improving steadily in their lungs during the winter, appeared to their friends to be getting worse, by reason of their constant fits of coughing and, in some cases, almost total loss of voice.

Many persons who were perfectly well in health and who were residing at Davos as visitors, complained of constant relaxed and sore throats, which could only be attributed to the dryness of the atmosphere. This was more especially the case in the dry and beautiful winter of 1879-80, and many of these sufferers were greatly relieved by a change to milder and damper weather. Invalids, on the other hand, who were suffering from diseased lungs of apparently every known form, but not complicated with affections of the bronchial tubes or throat, seemed, with some few exceptions, to improve steadily, and sometimes rapidly, at Davos. The exceptions were mostly those who, to a non-professional observer, seemed too far gone when they arrived to have much hope of recovery anywhere. Too feeble to take exercise, they shivered from morning to night, with their extremities blue and shrivelled by the lowness of the temperature, which seemed to be hardly counteracted in their cases by the warmth of the sun. These persons would certainly have spent a happier existence in warmer climes, though they might possibly have gained no great benefit from any change.

I must say, too, though with the greatest hesitation, as I believe it to be opposed to the opinion of the medical men in the place, that it seems to me doubtful how far the place is suitable to persons who are in that stage of consumption when hectic or recurrent fevers are high and frequent. I saw several cases amongst my own acquaintances in Davos of this fever, especially during my first winter; and they never seemed to be able to shake it off during the whole time they were there. Nearly all of these persons have since died; and it may be that their cases were almost hopeless from the first; but I think it right to mention my experience in this respect.

So far as I could make out from my own observation and from enquiries amongst the patients themselves, I can say, without hesitation, that nearly all the invalids who were suffering from diseases of the lungs, with the exceptions which I have before mentioned, seemed to derive real benefit from their residence at Davos during each winter that I spent there. Some young men, who had hereditary predisposition to the disease and who had broken down in health from confinement to business, made the most extraordinary and rapid progress, and left Davos with the medical assurance that the disease had been entirely arrested, and that they were to all intents and purposes cured. Their outward appearance and habits seemed to confirm this opinion, but I cannot say whether these favourable conditions have been permanent.

Others again seemed to make steady progress during their residence at Davos, and to leave the place in fairly good health; but they broke down again on their removal to lower regions, or during the miserable English summer of 1879. Some of these returned to Davos for the ensuing winter, and soon picked up their healthy appearance and habits again; but it remains to be seen whether these will last when they leave the place. Others again, in whom the disease was further advanced, seemed to make steady though somewhat slow progress towards recovery; and from being able to walk slowly, on their first arrival, on the level roads in front of their hotels, got by degrees to go up slight inclines, and even the steeper inclines of the wooded hills behind the village; and these might frequently be seen, towards the end of the winter, skating slowly on the ice-rink, or toboggining in a gentle manner down the easier paths near their hotels. Most of these intended to go a second or, if necessary, a third winter to Davos, when they had reasonable hopes of being restored to a fair state of health.

There were others again in whom the disease was still further advanced, who also made perceptible progress in the right direction. These would not advance so far as to walk much uphill, still less to skate or toboggin; but they gradually prolonged their walks on the level roads, and appeared to increase in strength and nervous power; and left Davos with much brighter hopes than when they arrived. These also would probably return again and again; and, even if they could never hope to be restored to their former strength, they could at least look forward to passing several months of each winter in comparative health and comfort.

I may add, that Dr. Ruedi told me that during each winter I was at Davos, he had found that about 82 per cent. of his patients had derived more or less benefit from their residence there; and a large proportion of those had made considerable progress. The average was nearly the same during each winter, though the weather during the first winter was rather exceptionally bad, and during the last winter was very exceptionally good in Davos.

One of the problems about Davos which is still unsolved is, whether the benefits which are in so many cases decidedly derived from a residence there are lasting; or whether they are not in most cases cancelled by a removal to lower and damper regions? There is evidence, I believe, in favour of both theories; and I cannot say which preponderates. It would seem to a non-professional man more probable that an improvement in health which is obtained in a cold and bracing climate should be permanent, than a similar improvement under more relaxing influences. Many, however, of the German invalids who gained great benefit at Davos, have found it impossible to keep their health in their own country; and they have established themselves at Davos, where they have opened shops and bazaars, and are able to live and do their work there in comfort.

The curative properties of Davos have had many causes assigned to them by writers on the subject. They have been attributed to the lightness and diminished pressure of the atmosphere in that high altitude; the absence of humidity in the air; the freedom from wind and dust; the unusual amount of solar heat; or to the marvellous purity of the atmosphere and its freedom from putrifying germs. The supporters of the last named hypothesis point to the fact that meat, when exposed to the air, dries and does not putrify; and they allege that external sores heal quickly instead of suppurating. Probably all these causes combine to bring about the favourable conditions which undoubtedly are found to exist at Davos.

I venture, however, to think that one of the most important causes of all is the number of days in every winter when invalids can be out in the open air for a greater or less number of hours together without fear of catching cold, if they will only use reasonable caution. Their appetites and digestions improve, and they thus gradually gain strength to make head against the disease which was rapidly undermining both. It is wonderful how many of the invalids, who could hardly look at the food of their own homes, manage to eat and digest the less appetising food which is put before them at the table d'hôte of their hotels. Complaints of the food were common enough; but nevertheless, the dishes went away empty, and the weights of the invalids increased.

The great drawback to all high Alpine health resorts is the difficulty of deciding where to go when the great snow-melting begins in April, and when it is generally considered that the climate becomes unhealthy for invalids. It still remains to be determined whether the effects of this snow-melting are as injurious as they are represented to be. It is probable that greater care would have to be taken by invalids against chills and catarrhs, when there is so much moisture in the air as naturally results from the melting away of the vast accumulation of snow in those high valleys and on the mountains which surround them on all sides; but it must be remembered that this snow-melting is gradual. It begins in most years before the middle of March, and a very considerable part of the snow in the valley of Davos has gone entirely by the end of the first week in April, which is the time of the general exodus. It no doubt often happens that more snow keeps falling during April and May, and thaws rapidly; and this makes the roads and paths muddy and unpleasant for walkers. But is it worse than the rainy weather which is almost sure to be met with at that time of year in lower regions? I have been told by invalids who have made a point of staying at Davos till the end of May during the past three or four winters, that they never experienced the slightest ill-effects from it. They felt no more susceptibility to colds than at other times of thaw during the winter; and, though the roads were sometimes almost impassable, the paths on the wooded slopes of the mountains, and many of the slopes themselves, were clear of snow, dry, and in good condition for walking. The sun was often hot, and the days brilliantly fine, and they could lie out on rugs under the fir trees for hours together with their books, and enjoy themselves. Many of these invalids, who were nearly all Germans, assured me that they benefited quite as much during the months of April and May, as they did during the earlier part of the season. Even if these good results could hardly be predicted in the majority of cases, it is a great question whether invalids would not at any rate be quite as well, if not better off, at Davos during those two proverbially treacherous months than in some lower halting place, where the spring rains, damp soil, and luxuriant vegetation might be productive of quite as much moisture in the atmosphere, and far more relaxing effects on the invalids, than during the snow-melting at Davos. The local doctors, hotel-keepers, and most of the leading inhabitants, unhesitatingly answer no. They may be perfectly right; but this seems to me to be a most important point for the consideration of thoroughly independent and competent witnesses. It must be admitted that the evidence of the local authorities cannot be said to be entirely unprejudiced. I do not suggest that it is intentionally prejudiced; but there has been a tradition of the unhealthiness of the snow-melting season from time immemorial; and I have not the slightest doubt that it is most honestly and firmly accepted as an article of faith, which it would be almost sacrilegious to question. It happens also to coincide with the interests of the believers. The doctors have been hard worked through the winter season, and will have to be ready for work again at the commencement of the summer season in June. It is most reasonable that they should wish to have a few weeks when they can get away, with their families, for a holiday. The hotel proprietors wish to have a regular cleaning up of their hotels during the slack season; and some of them, too, have no objection to a little change of air and scene at the same time. The bazaar keepers and tradesmen have the same natural desire for a holiday, and to take stock of their wares; and this can only be satisfactorily done at a time when there are comparatively few customers.

It thus comes to pass that there has been hitherto no one of importance left at Davos whose interest it is to raise any doubt as to the soundness of the tradition, except a very few invalids, who are not sorry to be left awhile in peace and quietness. This is, however, a question of the very highest importance to the great mass of invalids. Many of them would be thankful if they could remain quietly in their comfortable winter quarters until the season were sufficiently advanced for them to return to England, or to make a pleasant tour in other parts of the continent, instead of having to move at a bad time of the year to some other sanatorium, free indeed from the melting snow, but not improbably subject to other no less injurious conditions. It is certainly a fact that many of the more delicate invalids suffer from this change to a very considerable extent; and the question is, would they be likely to suffer less at Davos? This question ought to be carefully considered in any medical examination as to the merits of the place. If the answer were in the affirmative, then the value of Davos as a winter sanatorium would be enormously increased, and the place would be proportionately benefited; but if the answer were in the negative, then the further question arises, where should the invalids go for the months of April and May, so as to run the least risk of doing away with the good which they gained during the winter at Davos?

Three or four places are recommended by the local

physicians: Baden-Baden, Wiesbaden, Bern, Montreux on the Lake of Geneva, whence invalids can easily ascend to Glion rather later in the season, and Monte Generoso above the Lake of Lugano, which can be combined with a visit to the Italian lakes. I have been at all these places during the spring months, but I cannot offer any decided opinion as to their several qualifications. I may, however, shortly mention here some of what seem to me the advantages and disadvantages of each of these places.

Baden-Baden and Wiesbaden are both most beautifully situated, with numberless charming walks and drives in every direction. They are clean and well kept, and the water is good, and there are plenty of comfortable hotels in both places. The soil is sandy and dry, and the roads become quite dry in a wonderfully quick time after rain; and both places are surrounded by large forests of fir trees, which are considered to possess certain health-giving properties. They have also the great advantage of being but very little out of the direct route home to England. On the other hand, a great deal of rain falls over the whole district in which both places are situated, especially during the spring months. This is evidenced by the beautiful ferns and mosses and flowers which form a càrpet in all the woods, in spite of the dry nature of the soil, and by the full streams of water which pour down the slopes of the hills on all sides of the towns. In fine weather these places are apt to be hot and sultry, and in wet weather cold and chilly. I believe they are neither of them subject to very high or cold winds, being well protected by the hills and woods around them; but they are both decidedly relaxing places in spring; and I cannot conceive a greater change than from the bracing light air of Davos to the soft and muggy air of either of these places, which are both situated at the comparatively low altitude of from 500 to 600 feet above the level of the sea.

Bern, the capital of Switzerland, is an interesting and pleasant town, situated between 1,700 and 1,800 feet above the sea, with good hotels and *pensions*, and with pretty walks and drives in the neighbourhood, commanding in clear weather, beautiful views of the mountains of the Bernese Oberland. The town seems clean and healthy, but I should think it would be dull for a long residence. Two or three days always appear to me to be quite sufficient to exhaust the beauties of the place. Moreover, the winds blow keenly from the snow-covered mountains in the spring months, and the town is fully exposed to their force.

Montreux is situated near the head of the Lake of Geneva, between 1,200 and 1,300 feet above the sea level. It is well protected from the cold winds, and has a considerable reputation as a winter residence for invalids with chest complaints. I have, however, always found the comparatively low-lying districts of Switzerland wet and cold in the spring months, and I have no great liking for the shores of a large lake at that time of year.

Glion, which is about 1,000 feet above Montreux, is a charming place later in the season; but invalids ought not to go there before the end of May, or early in June; and it is not suitable for those who are unable to walk much uphill.

Monte Generoso is also a most charming place in June; but it is not fit for invalids before that time. The hotel is opened on the 1st of May, but does not profess to receive visitors before the 15th, and it would take a good month to get it properly aired for invalids after being shut up during the whole of the winter. The mountain is rather more than 5,500 feet in height, and the hotel is about 1,000 feet below the summit, and fairly sheltered from the cold winds. The views from all parts of the mountain are magnificent, and the walks are numerous and beautiful, whilst the wild flowers surpass in beauty and variety those of any other place which I ever visited. It is not, however, an easy place to get to by great invalids, as the ascent from Mendrisio, on the Lake of Lugano, is steep, and must be made on foot or on horseback; and the walks on the mountain are far more fitted for those who are sound than for the weak. Moreover, the mountain seems just the proper height for catching every cloud that passes through the sky; and visitors must be prepared for a great deal of rain, cloud, and mist, even when the whole district below them is bathed in sunshine. The soil, too, is deep and damp; and there is too often a feeling of dampness and chilliness around one, even in moderately fine weather. The Italian lakes are beautiful; but the weather in April and early May is too often wet and stormy to be healthful for invalids, and the climate is undoubtedly relaxing.

It is true that the weather during April and May is often very bad at Davos; but the invalid has always his warm hotel in which to remain on bad days; and I cannot feel sure that he would be much better off at any one of the places which I have here mentioned as being almost the only alternatives.

The only other places that I know of as being likely to be approximately suitable for the required purpose, are Bex or Aigle in the valley of the Rhone, both of which are situated at an elevation of about 1,400 feet above the sea, and in the midst of beautiful scenery. There are excellent hotels at both of these places, situated at some distance from, and at a considerable altitude above, the villages which are in the valley, and well above the mists which hang over the river in the early mornings and evenings, and which give the valley such an unenviable reputation for miasma. I believe, however, that, for some reason or other, neither of these places is strongly recommended by the medical men at Davos, though I should have preferred them to any of the places above mentioned.

Davos is not a difficult place to get to from England. It is distant less than thirty miles from Landquart, which is on the line of railway between Bâle and Coire, and about forty miles from Coire. Diligences run between each of those places and Davos daily, the one from Landquart in about seven hours, and the one from Coire in about eight. Landquart is the most suitable for invalids of these two starting places, as the diligence leaves Landquart at a more convenient hour; and during the winter, when the diligences have to be exchanged for sleighs, there is always a covered sleigh between Landquart and Davos. There is a fair Inn at Landquart, close to the railway station, and an excellent road ascends to Davos, through the charming valley of the Prättigau. Invalids who make the journey in the winter time must be amply provided against cold, especially in their legs and feet, during the drive from Landquart. They will probably find the Swiss railway carriages too hot and ill-ventilated, and it is a great change from them into a not very airtight covered sleigh.

The railway journey from England is not a very fatiguing one, especially if the invalid travel by way of Dover and Ostend; as there are carriages attached to the train at Ostend which convey passengers through to Bâle without a change. Invalids may rest at Bâle before going forward to Landquart, which is a slow journey of about seven hours.

Davos is, however, a difficult place for an invalid to get away from during the winter, if he find the place unsuitable for him. If he wish to go to more southern climes, he must return to Zurich, and travel thence by Bern and Geneva, before getting on to the main lines of railway to the south of France or to Italy. There are several diligence roads from Davos over the Alpine Passes into Italy, which would greatly shorten the journey; but the weather would be too cold in winter, and the roads too rough and dangerous from avalanches in spring, to admit of their being relied on as available. I venture, however, to think that, if only reasonable care and caution be exercised in the selection of cases suitable for a winter residence in Davos, there will be very few amongst those who go there who will wish to leave the place until the last day of the winter season.

I would ask in conclusion whether, if the evidence should turn out to be favourable to sending a large number of persons affected with lung disease to winter in high Alpine regions, there are not other places besides Davos which would be available for this purpose? Dr. Yeo says in his article that the Engadine has been tried and found wanting. No doubt the greater part of the valley of the Inn is eminently unsuitable for winter quarters; but has any fair trial been made of St. Moritz in modern times? The place is fairly protected from both cold and hot winds: and an hotel has been built and partly fitted up for winter invalids, and would soon be properly completed, if the place came more into repute. A gentleman, who has been there for many successive winters, has taken meteorological observations during his residence there, and has shown that the average cold is not much more intense than at Davos. His observations also tend to show that the winds there have not much greater force or frequency. I have very little doubt, from what I can learn from the most reliable sources, that Davos is, on the whole, the best and most advantageously situated of any of the high Alpine places that could be selected: but a time may come when it will be too much frequented by winter residents to be as healthful and agreeable as it is at present; and there will then be a general demand for some other and less crowded residence containing similar curative properties, and subject as nearly as possible, to the same favourable conditions as Davos.* It would seem advisable in the meantime to arrive at some satisfactory conclusions as to how this demand is to be supplied.

* Since writing the above I have received reliable information that this time has already come.

THE END.

50, ALBEMARLE STREET, LONDON.
January, 1882.

MR. MURRAY'S
GENERAL LIST OF WORKS.

ALBERT MEMORIAL. A Descriptive and Illustrated Account of the National Monument erected to the PRINCE CONSORT at Kensington. Illustrated by Engravings of its Architecture, Decorations, Sculptured Groups, Statues, Mosaics, Metalwork &c. With Descriptive Text. By DOYNE C. BELL. With 24 Plates. Folio. 12*l.* 12*s.*

——————— HANDBOOK. Fcap. 8vo. 1*s.*; or Illustrated Edition, 1*s.* 6*d.*

——————— (PRINCE) SPEECHES AND ADDRESSES. Fcap. 8vo. 1*s.*

ABBOTT (REV. J.). Memoirs of a Church of England Missionary in the North American Colonies. Post 8vo. 2*s.*

ABERCROMBIE (JOHN). Enquiries concerning the Intellectual Powers and the Investigation of Truth. Fcap. 8vo. 3*s.* 6*d.*

ACLAND (REV. CHARLES). Popular Account of the Manners and Customs of India. Post 8vo. 2*s.*

ÆSOP'S FABLES. A New Version. By Rev. THOMAS JAMES. With 100 Woodcuts, by TENNIEL and WOLF. Post 8vo. 2*s.* 6*d.*

AGRICULTURAL (ROYAL) JOURNAL. (*Published half-yearly.*)

AMBER-WITCH (THE). A most interesting Trial for Witchcraft. Translated by LADY DUFF GORDON. Post 8vo. 2*s.*

APOCRYPHA: With a Commentary Explanatory and Critical, by various Writers. Edited by REV. HENRY WACE. 2 Vols. Medium 8vo. Uniform with the Speaker's Commentary. [*In the Press.*

ARISTOTLE. [See GROTE, HATCH.]

ARMY LIST (THE). *Published Monthly by Authority.*

——————— (THE NEW OFFICIAL). *Published Quarterly.* Royal 8vo. 15*s.*

ARTHUR'S (LITTLE) History of England. By LADY CALLCOTT. *New Edition, continued to* 1878. With 36 Woodcuts. Fcap. 8vo. 1*s.* 6*d.*

ATKINSON (DR. R.) Vie de Seint Auban. A Poem in Norman-French. Ascribed to MATTHEW PARIS. With Concordance, Glossary and Notes. Small 4to. 10*s.* 6*d.*

AUSTIN (JOHN). LECTURES ON GENERAL JURISPRUDENCE; or, the Philosophy of Positive Law. Edited by ROBERT CAMPBELL. 2 Vols. 8vo. 32*s.*

——————— STUDENT'S EDITION, compiled from the above work, by ROBERT CAMPBELL. Post 8vo. 12*s.*

——————— Analysis of. By GORDON CAMPBELL. Post 8vo. 6*s.*

B

ADMIRALTY PUBLICATIONS; Issued by direction of the Lords Commissioners of the Admiralty:—
CHALLENGER EXPEDITION, 1873—1876: Report of the Scientific Results of. Zoology. Vol. I. 37s. 6d. Vol. II. 50s.
A MANUAL OF SCIENTIFIC ENQUIRY, for the Use of Travellers. *Fourth Edition.* Edited by ROBERT MAIN, M.A. Woodcuts. Post 8vo. 8s. 6d.
GREENWICH ASTRONOMICAL OBSERVATIONS, 1841 to 1847, and 1847 to 1877. Royal 4to. 20s. each.
GREENWICH ASTRONOMICAL RESULTS, 1847 to 1877. 4to. 3s. each.
MAGNETICAL AND METEOROLOGICAL OBSERVATIONS, 1844 to 1877. Royal 4to. 20s. each.
MAGNETICAL AND METEOROLOGICAL RESULTS, 1848 to 1877. 4to. 3s. each.
APPENDICES TO OBSERVATIONS.
   1837. Logarithms of Sines and Cosines in Time. 3s.
   1842. Catalogue of 1439 Stars, from Observations made in 1836. 1841. 4s.
   1845. Longitude of Valentia (Chronometrical). 3s.
   1847. Description of Altazimuth. 3s.
       Description of Photographic Apparatus. 2s.
   1851. Maskelyne's Ledger of Stars. 3s.
   1852. I. Description of the Transit Circle. 3s.
   1853. Bessel's Refraction Tables. 3s.
   1854. 1. Description of the Reflex Zenith Tube. 3s.
      II. Six Years' Catalogue of Stars, from Observations. 1848 to 1853. 4s.
   1860. Reduction of Deep Thermometer Observations. 2s.
   1862. II. Plan of Ground and Buildings of Royal Observatory, Greenwich. 3s.
      III. Longitude of Valentia (Galvanic). 2s.
   1864. I. Moon's Semi-diameter, from Occultations. 2s.
      II. Reductions of Planetary Observations. 1831 to 1835. 2s.
   1868. I. Corrections of Elements of Jupiter and Saturn. 2s.
      II. Second Seven Years' Catalogue of 2760 Stars. 1861-7. 4s.
      III. Description of the Great Equatorial. 3s.
   1871. Water Telescope. 3s.
   1873. Regulations of the Royal Observatory. 2s.
   1876. II. Nine Years' Catalogue of 2258 Stars. (1868-76.) 6s.
Cape of Good Hope Observations (Star Ledgers): 1856 to 1863. 2s.
——————————— 1856. 5s.
——————————— Astronomical Results. 1857 to 1858. 5s.
Cape Catalogue of 1159 Stars, reduced to the Epoch 1860. 3s.
Cape of Good Hope Astronomical Results. 1859 to 1860. 5s.
——————————— 1871 to 1873. 5s.
——————————— 1874 to 1876. 5s. each.
Report on Teneriffe Astronomical Experiment. 1856. 5s.
Paramatta Catalogue of 7385 Stars. 1822 to 1826. 4s.
REDUCTION OF THE OBSERVATIONS OF PLANETS. 1750 to 1830. Royal 4to. 20s. each.
——————————— LUNAR OBSERVATIONS. 1750 to 1830. 2 Vols. Royal 4to. 20s. each.
——————————— 1831 to 1851. 4to. 10s. each.
——————————— GREENWICH METEOROLOGICAL OBSERVATIONS. Chiefly 1847 to 1873. 5s.
ARCTIC PAPERS. 13s. 6d.
BERNOULLI'S SEXCENTENARY TABLE. 1779. 4to. 5s.
BESSEL'S AUXILIARY TABLES FOR HIS METHOD OF CLEARING LUNAR DISTANCES. 8vo. 2s.
ENCKE'S BERLINER JAHRBUCH, for 1830. *Berlin,* 1828. 8vo. 9s.
HANNYNGTON'S HAVERSINES.
HANSEN'S TABLES DE LA LUNE. 4to. 20s.
LAX'S TABLES FOR FINDING THE LATITUDE AND LONGITUDE. 1821. 8vo. 10s.

ADMIRALTY PUBLICATIONS—*continued.*
   LUNAR OBSERVATIONS at GREENWICH. 1783 to 1819. Compared with the Tables, 1821. 4to. 7s. 6d.
   MACLEAR ON LACAILLE'S ARC OF MERIDIAN. 2 Vols. 20s. each.
   MAYER'S DISTANCES of the MOON'S CENTRE from the PLANETS. 1822, 3s.; 1823, 4s. 6d. 1824 to 1835. 8vo. 4s. each.
   MAYER'S TABULÆ MOTUUM SOLIS ET LUNÆ. 1770. 5s.
   ———— ASTRONOMICAL OBSERVATIONS MADE AT GÖTTINGEN, from 1756 to 1761. 1826. Folio. 7s. 6d.
   NAUTICAL ALMANACS, from 1767 to 1884. 2s. 6d. each.
   ———————— SELECTIONS FROM, up to 1812. 8vo. 5s. 1834-54. 5s.
   ———————— SUPPLEMENTS, 1828 to 1833, 1837 and 1838. 2s. each.
   ———————— TABLE requisite to be used with the N.A. 1781. 8vo. 5s.
   SABINE'S PENDULUM EXPERIMENTS to DETERMINE THE FIGURE OF THE EARTH. 1825. 4to. 40s.
   SHEPHERD'S TABLES for CORRECTING LUNAR DISTANCES. 1772. Royal 4to. 21s.
   ———— TABLES, GENERAL, of the MOON'S DISTANCE from the SUN, and 10 STARS. 1787. Folio. 5s. 6d.
   TAYLOR'S SEXAGESIMAL TABLE. 1780. 4to. 15s.
   ———— TABLES OF LOGARITHMS. 4to. 60s.
   TIARK'S ASTRONOMICAL OBSERVATIONS for the LONGITUDE of MADEIRA. 1822. 4to. 5s.
   ———— CHRONOMETRICAL OBSERVATIONS for DIFFERENCES of LONGITUDE between DOVER, PORTSMOUTH, and FALMOUTH. 1823. 4to. 5s.
   VENUS and JUPITER: OBSERVATIONS of, compared with the TABLES. *London*, 1822. 4to. 2s.
   WALES AND BAYLY'S ASTRONOMICAL OBSERVATIONS. 1777. 4to. 21s.
   ———— REDUCTION OF ASTRONOMICAL OBSERVATIONS MADE IN THE SOUTHERN HEMISPHERE. 1764—1771. 1788. 4to. 10s. 6d.

BARBAULD (MRS.). Hymns in Prose for Children. With 100 Illustrations. 16mo. 3s. 6d.

BARCLAY (BISHOP). Extracts from the Talmud, illustrating the Teaching of the Bible. With an Introduction. 8vo. 14s.

BARKLEY (H. C.). Five Years among the Bulgarians and Turks between the Danube and the Black Sea. Post 8vo. 10s. 6d.

———— Bulgaria Before the War; during a Seven Years' Experience of European Turkey and its Inhabitants. Post 8vo. 10s. 6d.

———— My Boyhood: a True Story. Illustrations. Post 8vo. 6s.

BARROW (JOHN). Life, Exploits, and Voyages of Sir Francis Drake. Post 8vo. 2s.

BARRY (CANON). The Manifold Witness for Christ. An Attempt to Exhibit the Combined Force of various Evidences, Direct and Indirect of Christianity. 8vo. 12s.

———— (EDW. M.), R.A. Lectures on Architecture, delivered before the Royal Academy. Edited, with Memoir by Canon Barry. Portrait and Illustrations. 8vo. 16s.

BATES (H. W.). Records of a Naturalist on the Amazons during Eleven Years' Adventure and Travel. Illustrations. Post 8vo. 7s. 6d.

BAX (CAPT.). Russian Tartary, Eastern Siberia, China, Japan, &c. Illustrations. Cr. 8vo. 12s.

BELCHER (LADY). Account of the Mutineers of the 'Bounty, and their Descendants; with their Settlements in Pitcairn and Norfolk Islands. Illustrations. Post 8vo. 12s.

BELL (Sir Chas.). Familiar Letters. Portrait. Post 8vo. 12s.

—— (Doyne C.). Notices of the Historic Persons buried in the Chapel of St. Peter ad Vincula, in the Tower of London, with an account of the discovery of the supposed remains of Queen Anne Boleyn. With Illustrations. Crown 8vo. 14s.

BERTRAM (Jas. G.). Harvest of the Sea: an Account of British Food Fishes, including Fisheries and Fisher Folk. Illustrations. Post 8vo. 9s.

BIBLE COMMENTARY. The Old Testament. Explanatory and Critical. With a Revision of the Translation. By BISHOPS and CLERGY of the ANGLICAN CHURCH. Edited by F. C. Cook, M.A., Canon of Exeter. 6 Vols. Medium 8vo. 6l. 15s.

Vol. I. 30s. { Genesis. Exodus. Leviticus. Numbers. Deuteronomy. }

Vols. II. and III. 36s. { Joshua, Judges, Ruth, Samuel, Kings, Chronicles, Ezra, Nehemiah, Esther. }

Vol. IV. 24s. { Job. Psalms. Proverbs. Ecclesiastes. Song of Solomon. }

Vol. V. 20s. { Isaiah. Jeremiah. }

Vol. VI. 25s. { Ezekiel. Daniel. Minor Prophets. }

Bible Commentary. The New Testament. 4 Vols. Medium 8vo.

Vol. I. 18s. { Introduction. St. Matthew. St. Mark. St. Luke. }

Vol. II. 20s. { St. John. Acts. }

Vol. III. 28s. { Romans, Corinthians, Galatians, Philippians, Ephesians, Colossians, Thessalonians, Pastoral Epistles, Philemon. }

Vol. IV. 28s. { Hebrews, St. James, St. Peter, St. John, St. Jude, Revelation. }

—— The Student's Edition. Abridged and Edited by John M. Fuller, M.A., Vicar of Bexley. (To be completed in 6 Volumes.) Vols. I. to IV. Crown 8vo. 7s. 6d. each.

BIGG-WITHER (T. P.). Pioneering in South Brazil; Three Years of Forest and Prairie Life in the Province of Parana. Map and Illustrations. 2 vols. Crown 8vo. 24s.

BIRD (Isabella). Hawaiian Archipelago; or Six Months among the Palm Groves, Coral Reefs, and Volcanoes of the Sandwich Islands. Illustrations. Crown 8vo. 7s. 6d.

—— Unbeaten Tracks in Japan: Travels of a Lady in the interior, including Visits to the Aborigines of Yezo and the Shrines of Nikko and Ise. Illustrations. 2 Vols. Crown 8vo. 24s.

—— Lady's Life in the Rocky Mountains. Illustrations. Post 8vo. 7s. 6d.

BISSET (Sir John). Sport and War in South Africa from 1834 to 1867, with a Narrative of the Duke of Edinburgh's Visit. Illustrations. Crown 8vo. 14s.

BLUNT (Lady Anne). The Bedouins of the Euphrates Valley. With some account of the Arabs and their Horses. Illustrations. 2 Vols. Crown 8vo. 24s.

—— A Pilgrimage to Nejd, the Cradle of the Arab Race, and a Visit to the Court of the Arab Emir. Illustrations. 2 Vols. Post 8vo. 24s

BLUNT (Rev. J. J.). Undesigned Coincidences in the Writings of the Old and New Testaments, an Argument of their Veracity. Post 8vo. 6s.
—————— History of the Christian Church in the First Three Centuries. Post 8vo. 6s.
—————— Parish Priest; His Duties, Acquirements, and Obligations. Post 8vo. 6s.
—————— University Sermons. Post 8vo. 6s.

BOOK OF COMMON PRAYER. Illustrated with Coloured Borders, Initial Letters, and Woodcuts. 8vo. 18s.

BORROW (George). Bible in Spain; or the Journeys, Adventures, and Imprisonments of an Englishman in an Attempt to circulate the Scriptures in the Peninsula. Post 8vo. 5s.
—————— Gypsies of Spain; their Manners, Customs, Religion, and Language. With Portrait. Post 8vo. 5s.
—————— Lavengro; The Scholar—The Gypsy—and the Priest. Post 8vo. 5s.
—————— Romany Rye—a Sequel to "Lavengro." Post 8vo. 5s.
—————— Wild Wales: its People, Language, and Scenery. Post 8vo. 5s.
—————— Romano Lavo-Lil; Word-Book of the Romany, or English Gypsy Language; with Specimens of their Poetry, and an account of certain Gypsyries. Post 8vo. 10s. 6d.

BOSWELL'S Life of Samuel Johnson, LL.D. Including the Tour to the Hebrides. Edited by Mr. Croker. Seventh Edition. Portraits. 1 vol. Medium 8vo. 12s.

BRACE (C. L.). Manual of Ethnology; or the Races of the Old World. Post 8vo. 6s.

BREWER (Rev. J. S.). English Studies, or Essays on English History and Literature. 8vo. 14s. Contents:—

New Sources of English History.	The Stuarts.
Green's Short History of the English People.	Shakspeare.
Royal Supremacy.	Study of History and English History.
Hatfield House.	Ancient London.

BRITISH ASSOCIATION REPORTS. 8vo.
York and Oxford, 1831-32, 13s. 6d.
Cambridge, 1833, 12s.
Edinburgh, 1834, 15s.
Dublin, 1835, 13s. 6d.
Bristol, 1836, 12s.
Liverpool, 1837, 16s. 6d.
Newcastle, 1838, 15s.
Birmingham, 1839, 13s. 6d.
Glasgow, 1840, 15s.
Plymouth, 1841, 13s. 6d.
Manchester, 1842, 10s. 6d.
Cork, 1843, 12s.
York, 1844, 20s.
Cambridge, 1845, 12s.
Southampton, 1846, 15s.
Oxford, 1847, 18s.
Swansea, 1848, 9s.
Birmingham, 1849, 10s.
Edinburgh, 1850, 15s.
Ipswich, 1851, 16s. 6d.
Belfast, 1852, 15s.
Hull, 1853, 10s. 6d.
Liverpool, 1854, 18s.
Glasgow, 1855, 15s.
Cheltenham, 1856, 18s.
Dublin, 1857, 15s.
Leeds, 1858, 20s.
Aberdeen, 1859, 15s.
Oxford, 1860, 25s.
Manchester, 1861, 15s.
Cambridge, 1862, 20s.
Newcastle, 1863, 25s.
Bath, 1864, 18s.
Birmingham, 1865, 25s.
Nottingham, 1866, 24s.
Dundee, 1867, 26s.
Norwich, 1868, 25s.
Exeter, 1869, 22s.
Liverpool, 1870, 18s.
Edinburgh, 1871, 16s.
Brighton, 1872, 24s.
Bradford, 1873, 25s.
Belfast, 1874, 25s.
Bristol, 1875, 25s.
Glasgow, 1876, 25s.
Plymouth, 1877, 24s.
Dublin, 1878, 24s.
Sheffield, 1879, 24s.
Swansea, 1880, 24s.

BRUGSCH (Professor). A History of Egypt, under the Pharaohs. Derived entirely from Monuments, with a Memoir on the Exodus of the Israelites. New and revised Edition. Maps. 2 vols. 8vo. 32s.

BUNBURY (E. H.). A History of Ancient Geography, among the Greeks and Romans, from the Earliest Ages till the Fall of the Roman Empire. With Index and 20 Maps. 2 Vols. 8vo. 42s.

BURBIDGE (F. W.). The Gardens of the Sun: or A Naturalist's Journal on the Mountains and in the Forests and Swamps of Borneo and the Sulu Archipelago. With Illustrations. Crown 8vo. 14s.

BURCKHARDT'S Cicerone; or Art Guide to Painting in Italy. Translated from the German by Mrs. A. Clough. New Edition, revised by J. A. Crowe. Post 8vo. 6s.

BURN (Col.). Dictionary of Naval and Military Technical Terms, English and French—French and English. Crown 8vo. 15s.

BUTTMANN'S Lexilogus; a Critical Examination of the Meaning of numerous Greek Words, chiefly in Homer and Hesiod. By Rev. J. R. Fishlake. 8vo. 12s.

BUXTON (Charles). Memoirs of Sir Thomas Fowell Buxton, Bart. Portrait. 8vo. 16s. *Popular Edition*. Fcap. 8vo. 5s.

——— (Sydney C.). A Handbook to the Political Questions of the Day; with the Arguments on Either Side. 8vo. 6s.

BYLES (Sir John). Foundations of Religion in the Mind and Heart of Man. Post 8vo. 6s.

BYRON'S (Lord) LIFE AND WORKS:—

    Life, Letters, and Journals. By Thomas Moore. *Cabinet Edition*. Plates. 6 Vols. Fcap. 8vo. 18s.; or One Volume, Portraits. Royal 8vo., 7s. 6d.

    Life and Poetical Works. *Popular Edition*. Portraits. 2 vo's. Royal 8vo. 15s.

    Poetical Works. *Library Edition*. Portrait. 6 Vols. 8vo. 45s.

    Poetical Works. *Cabinet Edition*. Plates. 10 Vols. 12mo. 30s.

    Poetical Works. *Pocket Ed.* 8 Vols. 16mo. In a case. 21s.

    Poetical Works. *Popular Edition*. Plates. Royal 8vo. 7s. 6d.

    Poetical Works. *Pearl Edition*. Crown 8vo. 2s. 6d.

    Childe Harold. With 80 Engravings. Crown 8vo. 12s.

    Childe Harold. 16mo. 2s. 6d.

    Childe Harold. Vignettes. 16mo. 1s.

    Childe Harold. Portrait. 16mo. 6d.

    Tales and Poems. 16mo. 2s. 6d.

    Miscellaneous. 2 Vols. 16mo. 5s.

    Dramas and Plays. 2 Vols. 16mo. 5s.

    Don Juan and Beppo. 2 Vols. 16mo. 5s.

    Beauties. Poetry and Prose. Portrait. Fcap. 8vo. 3s. 6d.

CAMPBELL (Lord). Life: Based on his Autobiography, with selections from Journals, and Correspondence. Edited by Mrs. Hardcastle. Portrait. 2 Vols. 8vo. 30s.

——— Lord Chancellors and Keepers of the Great Seal of England. From the Earliest Times to the Death of Lord Eldon in 1838. 10 Vols. Crown 8vo. 6s. each.

——— Chief Justices of England. From the Norman Conquest to the Death of Lord Tenterden. 4 Vols. Crown 8vo. 6s. each.

——— (Thos.) Essay on English Poetry. With Short Lives of the British Poets. Post 8vo. 3s. 6d.

CARNARVON (LORD). Portugal, Gallicia, and the Basque Provinces. Post 8vo. 3s. 6d.

—————— The Agamemnon : Translated from Æschylus. Sm. 8vo. 6s.

CARNOTA (CONDE DA). Memoirs of the Life and Eventful Career of F.M. the Duke of Saldanha; Soldier and Statesman. With Selections from his Correspondence.* 2 Vols. 8vo. 32s.

CARTWRIGHT (W. C.). The Jesuits: their Constitution and Teaching. An Historical Sketch. 8vo. 9s.

CAVALCASELLE'S WORKS. [See CROWE.]

CESNOLA (GEN.). Cyprus; its Ancient Cities, Tombs, and Temples. Researches and Excavations during Ten Years' Residence in that Island. With 400 Illustrations. Medium 8vo. 50s.

CHILD (CHAPLIN). Benedicite; or, Song of the Three Children; being Illustrations of the Power, Beneficence, and Design manifested by the Creator in his Works. Post 8vo. 6s.

CHISHOLM (Mrs.). Perils of the Polar Seas; True Stories of Arctic Discovery and Adventure. Illustrations. Post 8vo. 6s.

CHURTON (ARCHDEACON). Poetical Remains, Translations and Imitations. Portrait. Post 8vo. 7s. 6d.

CLASSIC PREACHERS OF THE ENGLISH CHURCH. Being Lectures delivered at St. James', Westminster, in 1877-8. By Eminent Divines. With Introduction by J. E. Kempe. 2 Vols. Post 8vo. 7s. 6d. each.

CLIVE'S (LORD) Life. By REV. G. R. GLEIG. Post 8vo. 3s. 6d.

CLODE (C. M.). Military Forces of the Crown; their Administration and Government. 2 Vols. 8vo. 21s. each.

—————— Administration of Justice under Military and Martial Law, as applicable to the Army, Navy, Marine, and Auxiliary Forces. 8vo. 12s.

COLERIDGE'S (SAMUEL TAYLOR) Table-Talk. Portrait. 12mo. 3s. 6d.

COLONIAL LIBRARY. [See Home and Colonial Library.]

COMPANIONS FOR THE DEVOUT LIFE. Lectures on well-known Devotional Works. By Eminent Divines. With Preface by J. E. Kempe, M.A. Crown 8vo. 6s.

DE IMITATIONE CHRISTI.
PENSÉES OF BLAISE PASCAL.
S. FRANÇOIS DE SALES.
BAXTER'S SAINTS' REST.
S. AUGUSTINE'S CONFESSIONS.
TAYLOR'S HOLY LIVING AND DYING.
THEOLOGIA GERMANICA.
FÉNÉLON'S ŒUVRES SPIRITUELLES.
ANDREWES' DEVOTIONS.
CHRISTIAN YEAR.
PARADISE LOST.
PILGRIM'S PROGRESS.
PRAYER BOOK.

CONVOCATION PRAYER-BOOK. (See Prayer-Book.)

COOKE (E. W.). Leaves from my Sketch-Book. Being a Selection from Sketches made during many Tours. With Descriptive Text. 50 Plates. 2 Vols. Small folio. 31s. 6d. each.

COOKERY (MODERN DOMESTIC). Founded on Principles of Economy and Practical Knowledge, and Adapted for Private Families. By a Lady. Woodcuts. Fcap. 8vo. 5s.

CRABBE (REV. GEORGE). Life & Poetical Works. Illustrations. Royal 8vo. 7s.

CRIPPS (WILFRED). Old English Plate: Ecclesiastical, Decorative, and Domestic, its Makers and Marks. With a Complete Table of Date Letters, &c. New Edition. With 70 Illustrations. Medium 8vo. 16s.

—————— Old French Plate; Furnishing Tables of the Paris Date Letters, and Facsimiles of Other Marks. With Illustrations. 8vo. 8s. 6d.

CROKER (J. W.). Progressive Geography for Children. 18mo. 1s. 6d.
—— Boswell's Life of Johnson. Including the Tour to the Hebrides. *Seventh Edition.* Portraits. 8vo. 12s.
—— Historical Essay on the Guillotine. Fcap. 8vo. 1s.
CROWE AND CAVALCASELLE. Lives of the Early Flemish Painters. Woodcuts. Post 8vo, 7s. 6d.; or Large Paper, 8vo, 15s.
—— History of Painting in North Italy, from 14th to 16th Century. With Illustrations. 2 Vols. 8vo. 42s.
—— Life and Times of Titian, with some Account of his Family, chiefly from new and unpublished records. With Portrait and Illustrations. 2 vols. 8vo. 21s.
CUMMING (R. GORDON). Five Years of a Hunter's Life in the Far Interior of South Africa. Woodcuts. Post 8vo. 6s.
CUNYNGHAME (SIR ARTHUR). Travels in the Eastern Caucasus, on the Caspian and Black Seas, in Daghestan and the Frontiers of Persia and Turkey. Illustrations. 8vo. 18s.
CURTIUS' (PROFESSOR) Student's Greek Grammar, for the Upper Forms. Edited by DR. WM. SMITH. Post 8vo. 6s.
—— Elucidations of the above Grammar. Translated by EVELYN ABBOT. Post 8vo. 7s. 6d.
—— Smaller Greek Grammar for the Middle and Lower Forms. Abridged from the larger work. 12mo. 3s. 6d.
—— Accidence of the Greek Language. Extracted from the above work. 12mo. 2s. 6d.
—— Principles of Greek Etymology. Translated by A. S. WILKINS, M.A., and E. B. ENGLAND, M.A. 2 vols. 8vo. 15s. each.
—— The Greek Verb, its Structure and Development. Translated by A. S. WILKINS, M.A., and E. B. ENGLAND, M.A. 8vo. 18s.
CURZON (HON. ROBERT). Visits to the Monasteries of the Levant. Illustrations. Post 8vo. 7s. 6d.
CUST (GENERAL). Warriors of the 17th Century—The Thirty Years' War. 2 Vols. 16s. Civil Wars of France and England. 2 Vols. 16s. Commanders of Fleets and Armies. 2 Vols. 18s.
—— Annals of the Wars—18th & 19th Century. With Maps. 9 Vols. Post 8vo. 5s. each.
DAVY (SIR HUMPHRY). Consolations in Travel; or, Last Days of a Philosopher. Woodcuts. Fcap. 8vo. 3s. 6d.
—— Salmonia; or, Days of Fly Fishing. Woodcuts. Fcap. 8vo. 3s. 6d.
DE COSSON (E. A.). The Cradle of the Blue Nile; a Journey through Abyssinia and Soudan, and a Residence at the Court of King John of Ethiopia. Map and Illustrations. 2 vols. Post 8vo. 21s.
DENNIS (GEORGE). The Cities and Cemeteries of Etruria. A new Edition, revised, recording all the latest Discoveries. With 20 Plans and 200 Illustrations. 2 vols. Medium 8vo. 42s.
DENT (EMMA). Annals of Winchcombe and Sudeley. With 120 Portraits, Plates and Woodcuts. 4to. 42s.
DERBY (EARL OF). Iliad of Homer rendered into English Blank Verse. With Portrait. 2 Vols. Post 8vo. 10s.

DARWIN'S (CHARLES) WORKS:—
> JOURNAL OF A NATURALIST DURING A VOYAGE ROUND THE WORLD. Crown 8vo. 9s.
>
> ORIGIN OF SPECIES BY MEANS OF NATURAL SELECTION; or, the Preservation of Favoured Races in the Struggle for Life. Woodcuts. Crown 8vo. 7s. 6d.
>
> VARIATION OF ANIMALS AND PLANTS UNDER DOMESTICATION Woodcuts. 2 Vols. Crown 8vo. 18s.
>
> DESCENT OF MAN, AND SELECTION IN RELATION TO SEX. Woodcuts. Crown 8vo. 9s.
>
> EXPRESSIONS OF THE EMOTIONS IN MAN AND ANIMALS. With Illustrations. Crown 8vo. 12s.
>
> VARIOUS CONTRIVANCES BY WHICH ORCHIDS ARE FERTILIZED BY INSECTS. Woodcuts. Crown 8vo. 9s.
>
> MOVEMENTS AND HABITS OF CLIMBING PLANTS. Woodcuts, Crown 8vo. 6s.
>
> INSECTIVOROUS PLANTS. Woodcuts. Crown 8vo. 14s.
>
> EFFECTS OF CROSS AND SELF-FERTILIZATION IN THE VEGETABLE KINGDOM. Crown 8vo. 12s.
>
> DIFFERENT FORMS OF FLOWERS ON PLANTS OF THE SAME SPECIES. Crown 8vo. 10s. 6d.
>
> POWER OF MOVEMENT IN PLANTS. Woodcuts. Cr. 8vo. 15s.
>
> THE FORMATION OF VEGETABLE MOULD THROUGH THE ACTION OF WORMS; with Observations on their Habits. Post 8vo. 9s.
>
> LIFE OF ERASMUS DARWIN. With a Study of his Works by ERNEST KRAUSE. Portrait. Crown 8vo. 7s. 6d.
>
> FACTS AND ARGUMENTS FOR DARWIN. By FRITZ MULLER. Translated by W. S. DALLAS. Woodcuts. Post 8vo. 6s.

DERRY (BISHOP OF). Witness of the Psalms to Christ and Christianity. The Bampton Lectures for 1876. 8vo. 14s.

DEUTSCH (EMANUEL). Talmud, Islam, The Targums and other Literary Remains. With a brief Memoir. 8vo. 12s.

DILKE (SIR C. W.). Papers of a Critic. Selected from the Writings of the late CHAS. WENTWORTH DILKE. With a Biographical Sketch. 2 Vols. 8vo. 24s.

DOG-BREAKING. [See HUTCHINSON.]

DOUGLAS'S (SIR HOWARD) Theory and Practice of Gunnery. Plates. 8vo. 21s.

―――――― (WM.) Horse-Shoeing; As it Is, and As it Should be. Illustrations. Post 8vo. 7s. 6d.

DRAKE'S (SIR FRANCIS) Life, Voyages, and Exploits, by Sea and Land. By JOHN BARROW. Post 8vo. 2s.

DRINKWATER (JOHN). History of the Siege of Gibraltar, 1779-1783. With a Description and Account of that Garrison from the Earliest Periods. Post 8vo. 2s.

DUCANGE'S MEDIÆVAL LATIN-ENGLISH DICTIONARY. Re-arranged and Edited, in accordance with the modern Science of Philology, by Rev. E. A. DAYMAN and J. H. HESSELS. Small 4to. [*In Preparation.*

DU CHAILLU (PAUL B.). Land of the Midnight Sun; Summer and Winter Journeys through Sweden, Norway, Lapland, and Northern Finland, with Descriptions of the Inner Life of the People, their Manners and Customs, the Primitive Antiquities, &c., &c. With Map and 235 Illustrations. 2 Vols. 8vo. 36s.
———— Equatorial Africa, with Accounts of the Gorilla, the Nest-building Ape, Chimpanzee, Crocodile, &c. Illustrations. 8vo. 21s.
———— Journey to Ashango Land; and Further Penetration into Equatorial Africa. Illustrations. 8vo. 21s.
DUFFERIN (LORD). Letters from High Latitudes; a Yacht Voyage to Iceland, Jan Mayen, and Spitzbergen. Woodcuts. Post 8vo. 7s. 6d.
———— Speeches and Addresses, Political and Literary, delivered in the House of Lords, in Canada, and elsewhere. 8vo.
DUNCAN (MAJOR). History of the Royal Artillery. Compiled from the Original Records. Portraits. 2 Vols. 8vo. 18s.
———— English in Spain; or, The Story of the War of Succession, 1834-1840. Compiled from the Reports of the British Commissioners. With Illustrations. 8vo. 16s.
DÜRER (ALBERT); A History of his Life and Works. By MORIZ THAUSING. Translated from the German. Edited by FREDERICK A. EATON, Secretary of the Royal Academy. With Portrait and Illustrations. 2 vols. Medium 8vo. 42s.
EASTLAKE (SIR CHARLES). Contributions to the Literature of the Fine Arts. With Memoir of the Author, and Selections from his Correspondence. By LADY EASTLAKE. 2 Vols. 8vo. 24s.
EDWARDS (W. H.). Voyage up the River Amazon, including a Visit to Para. Post 8vo. 2s.
ELDON'S (LORD) Public and Private Life, with Selections from his Diaries, &c. By HORACE TWISS. Portrait. 2 Vols. Post 8vo. 21s
ELGIN (LORD). Letters and Journals. Edited by THEODORE WALROND. With Preface by Dean Stanley. 8vo. 14s.
ELLESMERE (LORD). Two Sieges of Vienna by the Turks. Translated from the German. Post 8vo. 2s.
ELLIS (W.). Madagascar Revisited. The Persecutions and Heroic Sufferings of the Native Christians. Illustrations. 8vo. 16s.
———— Memoir. By His Son. With his Character and Work. By REV. HENRY ALLON, D.D. Portrait. 8vo. 10s. 6d.
———— (ROBINSON) Poems and Fragments of Catullus. 16mo. 5s.
ELPHINSTONE (HON. MOUNTSTUART). History of India—the Hindoo and Mahomedan Periods. Edited by PROFESSOR COWELL. Map. 8vo. 18s.
———— (H. W.). Patterns for Turning; Comprising Elliptical and other Figures cut on the Lathe without the use of any Ornamental Chuck. With 70 Illustrations. Small 4to. 15s.
ELTON (CAPT.) and H. B. COTTERILL. Adventures and Discoveries Among the Lakes and Mountains of Eastern and Central Africa. With Map and Illustrations. 8vo. 21s.
ENGLAND. [See ARTHUR, CROKER, HUME, MARKHAM, SMITH, and STANHOPE.]
ESSAYS ON CATHEDRALS. Edited, with an Introduction. By DEAN HOWSON. 8vo. 12s.
FERGUSSON (JAMES). History of Architecture in all Countries from the Earliest Times. With 1,600 Illustrations. 4 Vols. Medium 8vo.
I. & II. Ancient and Mediæval. 63s.
III. Indian & Eastern. 42s. IV. Modern. 31s. 6d.

FERGUSSON (James). Rude Stone Monuments in all Countries;
their Age and Uses. With 230 Illustrations. Medium 8vo. 2ls.
——————— Holy Sepulchre and the Temple at Jerusalem.
Woodcuts. 8vo. 7s. 6d.
——————— Temples of the Jews and other buildings in
the Haram Area at Jerusalem. With Illustrations. 4to. 42s.
FLEMING (Professor). Student's Manual of Moral Philosophy.
With Quotations and References. Post 8vo. 7s. 6d.
FLOWER GARDEN. By Rev. Thos. James. Fcap. 8vo. 1s.
FORBES (Capt.) British Burma and its People; Native
Manners, Customs, and Religion. Cr. 8vo. 10s. 6d.
——————— (George). Electricity and its applications, as illustrated
by the Paris Exhibition of Electricity, 1881. Reprinted, with additions,
from the 'Times.' Post 8vo. [In the Press
FORD (Richard). Gatherings from Spain. Post 8vo. 3s. 6d.
FORSTER (John). The Early Life of Jonathan Swift. 1667-1711.
With Portrait. 8vo. 15s.
FORSYTH (William). Hortensius; an Historical Essay on the
Office and Duties of an Advocate. Illustrations. 8vo. 7s. 6d.
——————— Novels and Novelists of the 18th Century, in
Illustration of the Manners and Morals of the Age. Post 8vo. 10s. 6d.
FRANCE (History of). [See Jervis—Markham—Smith—Students'—Tocqueville.]
FRENCH IN ALGIERS; The Soldier of the Foreign Legion—
and the Prisoners of Abd-el-Kadir. Translated by Lady Duff Gordon.
Post 8vo. 2s.
FRERE (Sir Bartle). Indian Missions. Small 8vo. 2s. 6d.
——————— Eastern Africa as a Field for Missionary Labour. With
Map. Crown 8vo. 5s.
——————— Bengal Famine. How it will be Met and How to
Prevent Future Famines in India. With Maps. Crown 8vo. 5s.
——————— (M.). Old Deccan Days, or Hindoo Fairy Legends
current in Southern India, with Introduction by Sir Bartle Frere.
With 50 Illustrations. Post 8vo. 7s. 6d.
GALTON (F.). Art of Travel; or, Hints on the Shifts and Contrivances available in Wild Countries. Woodcuts. Post 8vo. 7s. 6d.
GEOGRAPHY. [See Bunbury—Croker—Richardson—Smith—Students'.]
GEOGRAPHICAL SOCIETY'S JOURNAL. (Published Yearly.)
GEORGE (Ernest). The Mosel; a Series of Twenty Etchings, with
Descriptive Letterpress. Imperial 4to. 42s.
——————— Loire and South of France; a Series of Twenty
Etchings, with Descriptive Text. Folio. 42s.
GERMANY (History of). [See Markham.]
GIBBON (Edward). History of the Decline and Fall of the
Roman Empire. Edited by Milman, Guizot, and Dr. Wm. Smith.
Maps. 8 Vols. 8vo. 60s.
——————— The Student's Edition; an Epitome of the above
work, incorporating the Researches of Recent Commentators. By Dr.
Wm. Smith. Woodcuts. Post 8vo. 7s. 6d.
GIFFARD (Edward). Deeds of Naval Daring; or, Anecdotes of
the British Navy. Fcap. 8vo. 3s. 6d.

GILL (CAPT. WM). The River of Golden Sand. Narrative of a Journey through China to Burmah. With a Preface by Col. H. Yule, C.B. Maps and Illustrations. 2 Vols. 8vo. 30s.

―――― (MRS.). Six Months in Ascension. An Unscientific Account of a Scientific Expedition. Map. Crown 8vo. 9s.

GLADSTONE (W. E.). Rome and the Newest Fashions in Religion. Three Tracts. 8vo. 7s. 6d.

―――――――― Gleanings of Past Years, 1843-78. 7 vols. Small 8vo. 2s. 6d. each. I. The Throne, the Prince Consort, the Cabinet and Constitution. II. Personal and Literary. III. Historical and Speculative. IV. Foreign. V. and VI. Ecclesiastical. VII. Miscellaneous.

GLEIG (G. R.). Campaigns of the British Army at Washington and New Orleans. Post 8vo. 2s.

―――― Story of the Battle of Waterloo. Post 8vo. 3s. 6d.

―――― Narrative of Sale's Brigade in Affghanistan. Post 8vo. 2s.

―――― Life of Lord Clive. Post 8vo. 3s. 6d.

―――――――― Sir Thomas Munro. Post 8vo. 3s. 6d.

GLYNNE (SIR STEPHEN R.). Notes on the Churches of Kent: With Preface by W. H. Gladstone, M.P. Illustrations. 8vo. 12s.

GOLDSMITH'S (OLIVER) Works. Edited with Notes by PETER CUNNINGHAM. Vignettes. 4 Vols. 8vo. 30s.

GOMM (SIR WM. M.), Commander-in-Chief in India, Constable of the Tower, and Colonel of the Coldstream Guards. His Letters and Journals. 1799 to 1815. Edited by F. C. Carr Gomm. With Portrait. 8vo. 12s.

GORDON (SIR ALEX.). Sketches of German Life, and Scenes from the War of Liberation. Post 8vo. 3s. 6d.

―――― (LADY DUFF) Amber-Witch: A Trial for Witchcraft. Post 8vo. 2s.

―――― French in Algiers. 1. The Soldier of the Foreign Legion. 2. The Prisoners of Abd-el-Kadir. Post 8vo. 2s.

GRAMMARS. [See CURTIUS; HALL; HUTTON; KING EDWARD; LEATHES; MAETZNER; MATTHIÆ; SMITH.]

GREECE (HISTORY OF). [See GROTE—SMITH—STUDENTS'.]

GROTE'S (GEORGE) WORKS:—

HISTORY OF GREECE. From the Earliest Times to the close of the generation contemporary with the Death of Alexander the Great. *Library Edition.* Portrait, Maps, and Plans. 10 Vols. 8vo. 120s. *Cabinet Edition.* Portrait and Plans. 12 Vols. Post 8vo. 6s. each.

PLATO, and other Companions of Socrates. 3 Vols. 8vo. 45s.

ARISTOTLE. With additional Essays. 8vo. 18s.

MINOR WORKS. Portrait. 8vo. 14s.

LETTERS ON SWITZERLAND IN 1847. 6s.

PERSONAL LIFE. Portrait. 8vo. 12s.

GROTE (MRS.). A Sketch. By LADY EASTLAKE. Crown 8vo. 6s.

HALL'S (T. D.) School Manual of English Grammar. With Copious Exercises. 12mo. 3s. 6d.

―――― Manual of English Composition. With Copious Illustrations and Practical Exercises. 12mo. 3s. 6d.

―――― Primary English Grammar for Elementary Schools. Based on the larger work. 16mo. 1s.

―――― Child's First Latin Book, comprising a full Practice of Nouns, Pronouns, and Adjectives, with the Active Verbs. 16mo.

HALLAM'S (HENRY) WORKS:—
 THE CONSTITUTIONAL HISTORY OF ENGLAND, from the Accession of Henry the Seventh to the Death of George the Second. *Library Edition*, 3 Vols. 8vo. 30s. *Cabinet Edition*, 3 Vols. Post 8vo. 12s. *Student's Edition*, Post 8vo. 7s. 6d.
 HISTORY OF EUROPE DURING THE MIDDLE AGES. *Library Edition*, 3 Vols. 8vo. 30s. *Cabinet Edition*, 3 Vols. Post 8vo. 12s. *Student's Edition*, Post 8vo. 7s. 6d.
 LITERARY HISTORY OF EUROPE DURING THE 15TH, 16TH, AND 17TH CENTURIES. *Library Edition*, 3 Vols. 8vo. 36s. *Cabinet Edition*, 4 Vols. Post 8vo. 16s.
 ———— (ARTHUR) Literary Remains; in Verse and Prose. Portrait. Fcap. 8vo. 3s. 6d.

HAMILTON (ANDREW). Rheinsberg: Memorials of Frederick the Great and Prince Henry of Prussia. 2 Vols. Crown 8vo. 21s.

HART'S ARMY LIST. (*Published Quarterly and Annually.*)

HATCH (W. M.). The Moral Philosophy of Aristotle, consisting of a translation of the Nichomachean Ethics, and of the Paraphrase attributed to Andronicus, with an Introductory Analysis of each book. 8vo. 18s.

HATHERLEY (LORD). The Continuity of Scripture, as Declared by the Testimony of our Lord and of the Evangelists and Apostles. 8vo. 6s. *Popular Edition*. Post 8vo. 2s. 6d.

HAY (SIR J. H. DRUMMOND). Western Barbary, its Wild Tribes and Savage Animals. Post 8vo. 2s.

HAYWARD (A.). Sketches of Eminent Statesmen and Writers, with other Essays. Reprinted from the "Quarterly Review." Contents: Thiers, Bismarck, Cavour, Metternich, Montalembert, Melbourne, Wellesley, Byron and Tennyson, Venice, St. Simon, Sevigné, Du Deffand, Holland House, Strawberry Hill. 2 Vols. 8vo. 28s.

HEAD'S (SIR FRANCIS) WORKS:—
 THE ROYAL ENGINEER. Illustrations. 8vo. 12s.
 LIFE OF SIR JOHN BURGOYNE. Post 8vo. 1s.
 RAPID JOURNEYS ACROSS THE PAMPAS. Post 8vo. 2s.
 BUBBLES FROM THE BRUNNEN OF NASSAU. Illustrations. Post 8vo. 7s. 6d.
 STOKERS AND POKERS; or, the London and North Western Railway. Post 8vo. 2s.

HEBER'S (BISHOP) Journals in India. 2 Vols. Post 8vo. 7s.
———— Poetical Works. Portrait. Fcap. 8vo. 3s. 6d.
———— Hymns adapted to the Church Service. 16mo. 1s. 6d.

HERODOTUS. A New English Version. Edited, with Notes and Essays, Historical, Ethnographical, and Geographical, by CANON RAWLINSON, SIR H. RAWLINSON and SIR J. G. WILKINSON. Maps and Woodcuts. 4 Vols. 8vo. 48s.

HERRIES (RT. HON. JOHN). Memoir of his Public Life during the Reigns of George III. and IV., William IV., and Queen Victoria. Founded on his Letters and other Unpublished Documents. By his son, Edward Herries, C.B. 2 vols. 8vo. 24s.

HERSCHEL'S (CAROLINE) Memoir and Correspondence. By MRS. JOHN HERSCHEL. With Portraits. Crown 8vo. 7s. 6d.

FOREIGN HANDBOOKS.

HAND-BOOK—TRAVEL-TALK. English, French, German, and Italian. 18mo. 3s. 6d.

————— HOLLAND AND BELGIUM. Map and Plans. Post 8vo. 6s.

————— NORTH GERMANY and THE RHINE,— The Black Forest, the Hartz, Thüringerwald, Saxon Switzerland, Rügen, the Giant Mountains, Taunus, Odenwald, Elass, and Lothringen. Map and Plans. Post 8vo. 10s.

————— SOUTH GERMANY,— Wurtemburg, Bavaria, Austria, Styria, Salzburg, the Alps, Tyrol, Hungary, and the Danube, from Ulm to the Black Sea. Maps and Plans. Post 8vo. 10s.

————— PAINTING. German, Flemish, and Dutch Schools. Illustrations. 2 Vols. Post 8vo. 24s.

————— LIVES AND WORKS OF EARLY FLEMISH Painters. Illustrations. Post 8vo. 7s. 6d.

————— SWITZERLAND, Alps of Savoy, and Piedmont. In Two Parts. Maps and Plans. Post 8vo. 10s.

————— FRANCE, Part I. Normandy, Brittany, the French Alps, the Loire, Seine, Garonne, and Pyrenees. Maps and Plans. Post 8vo. 7s. 6d.

————— Part II. Central France, Auvergne, the Cevennes, Burgundy, the Rhone and Saone, Provence, Nimes, Arles, Marseilles, the French Alps, Alsace, Lorraine, Champagne, &c. Maps and Plans. Post 8vo. 7s. 6d.

————— MEDITERRANEAN—its Principal Islands, Cities, Seaports, Harbours, and Border Lands. For travellers and yachtsmen, with nearly 50 Maps and Plans. Post 8vo. 20s.

————— ALGERIA AND TUNIS. Algiers, Constantine, Oran, the Atlas Range. Maps and Plans. Post 8vo. 10s.

————— PARIS, and its Environs. Maps and Plans. 16mo. 3s. 6d.

————— SPAIN, Madrid, The Castiles, The Basque Provinces, Leon, The Asturias, Galicia, Estremadura, Andalusia, Ronda, Granada, Murcia, Valencia, Catalonia, Aragon, Navarre, The Balearic Islands, &c. &c. Maps and Plans. Post 8vo. 20s.

————— PORTUGAL, Lisbon, Porto, Cintra, Mafra, &c. Map and Plan. Post 8vo. 12s.

————— NORTH ITALY, Turin, Milan, Cremona, the Italian Lakes, Bergamo, Brescia, Verona, Mantua, Vicenza, Padua, Ferrara, Bologna, Ravenna, Rimini, Piacenza, Genoa, the Riviera, Venice, Parma, Modena, and Romagna. Maps and Plans. Post 8vo. 10s.

————— CENTRAL ITALY, Florence, Lucca, Tuscany, The Marches, Umbria, &c. Maps and Plans. Post 8vo. 10s.

————— ROME AND ITS ENVIRONS. With 50 Maps and Plans. Post 8vo. 10s.

————— SOUTH ITALY, Naples, Pompeii, Herculaneum, and Vesuvius. Maps and Plans. Post 8vo. 10s.

————— PAINTING. The Italian Schools. Illustrations. 2 Vols. Post 8vo. 30s.

————— LIVES OF ITALIAN PAINTERS, FROM CIMABUE to BASSANO. By Mrs. JAMESON. Portraits. Post 8vo. 12s.

————— NORWAY, Christiania, Bergen, Trondhjem. The Fjelds and Fjords. Maps and Plans. Post 8vo. 9s.

————— SWEDEN, Stockholm, Upsala, Gothenburg, the Shores of the Baltic, &c. Maps and Plan. Post 8vo. 6s.

HAND-BOOK—DENMARK, Sleswig, Holstein, Copenhagen, Jutland, Iceland. Maps and Plans. Post 8vo. 6s.
———— RUSSIA, St. Petersburg, Moscow, Poland, and Finland. Maps and Plans. Post 8vo. 18s.
———— GREECE, the Ionian Islands, Continental Greece, Athens, the Peloponnesus, the Islands of the Ægean Sea, Albania, Thessaly, and Macedonia. Maps, Plans, and Views. Post 8vo.
———— TURKEY IN ASIA—Constantinople, the Bosphorus, Dardanelles, Broussa, Plain of Troy, Crete, Cyprus, Smyrna, Ephesus, the Seven Churches, Coasts of the Black Sea, Armenia, Euphrates Valley, Route to India, &c. Maps and Plans. Post 8vo. 15s.
———— EGYPT, including Descriptions of the Course of the Nile through Egypt and Nubia, Alexandria, Cairo, and Thebes, the Suez Canal, the Pyramids, the Peninsula of Sinai, the Oases, the Fyoom, &c. In Two Parts. Maps and Plans. Post 8vo. 15s.
———— HOLY LAND—Syria, Palestine, Peninsula of Sinai, Edom, Syrian Deserts, Petra, Damascus; and Palmyra. Maps and Plans. Post 8vo. 20s. *⁎* Travelling Map of Palestine. In a case. 12s.
———— INDIA. Maps and Plans. Post 8vo. Part I. Bombay, 15s. Part II. Madras, 15s. Part III. Bengal.

ENGLISH HAND-BOOKS.

HAND-BOOK—ENGLAND AND WALES. An Alphabetical Hand-Book. Condensed into One Volume for the Use of Travellers. With a Map. Post 8vo. 10s.
———— MODERN LONDON. Maps and Plans. 16mo. 3s. 6d.
———— ENVIRONS OF LONDON within a circuit of 20 miles. 2 Vols. Crown 8vo. 21s.
———— ST. PAUL'S CATHEDRAL. 20 Illustrations. Crown 8vo 10s. 6d.
———— EASTERN COUNTIES, Chelmsford, Harwich, Colchester, Maldon, Cambridge, Ely, Newmarket, Bury St. Edmunds, Ipswich, Woodbridge, Felixstowe, Lowestoft, Norwich, Yarmouth, Cromer, &c. Map and Plans. Post 8vo. 12s.
———— CATHEDRALS of Oxford, Peterborough, Norwich, Ely, and Lincoln. With 90 Illustrations. Crown 8vo. 21s.
———— KENT, Canterbury, Dover, Ramsgate, Sheerness, Rochester, Chatham, Woolwich. Maps and Plans. Post 8vo. 7s. 6d.
———— SUSSEX, Brighton, Chichester, Worthing, Hastings, Lewes, Arundel, &c. Maps and Plans. Post 8vo. 6s.
———— SURREY AND HANTS, Kingston, Croydon, Reigate, Guildford, Dorking, Boxhill, Winchester, Southampton, New Forest, Portsmouth, Isle of Wight, &c. Maps and Plans. Post 8vo. 10s.
———— BERKS, BUCKS, AND OXON, Windsor, Eton, Reading, Aylesbury, Uxbridge, Wycombe, Henley, the City and University of Oxford, Blenheim, and the Descent of the Thames. Maps and Plans. Post 8vo.
———— WILTS, DORSET, AND SOMERSET, Salisbury, Chippenham, Weymouth, Sherborne, Wells, Bath, Bristol, Taunton, &c. Map. Post 8vo. 10s.
———— DEVON, Exeter, Ilfracombe, Linton, Sidmouth, Dawlish, Teignmouth, Plymouth, Devonport, Torquay. Maps and Plans. Post 8vo. 7s. 6d.

HAND-BOOK—CORNWALL, Launceston, Penzance, Falmouth
the Lizard, Land's End, &c. Maps. Post 8vo. 6s.
————————CATHEDRALS of Winchester, Salisbury, Exeter
Wells, Chichester, Rochester, Canterbury, and St. Albans. With 130
Illustrations. 2 Vols. Cr. 8vo. 36s. St. Albans separately, cr. 8vo.
6s.
———————— GLOUCESTER, HEREFORD, AND WORCESTER,
Cirencester, Cheltenham, Stroud, Tewkesbury, Leominster, Ross, Malvern, Kidderminster, Dudley, Bromsgrove, Evesham. Map. Post 8vo.
———————— CATHEDRALS of Bristol, Gloucester, Hereford,
Worcester, and Lichfield. With 50 Illustrations. Crown 8vo. 16s.
———————— NORTH WALES, Bangor, Carnarvon, Beaumaris,
Snowdon, Llanberis, Dolgelly, Cader Idris, Conway, &c. Map. Post
8vo. 7s.
———————— SOUTH WALES, Monmouth, Llandaff, Merthyr,
Vale of Neath, Pembroke, Carmarthen, Tenby, Swansea, The Wye, &c.
Map. Post 8vo. 7s.
———————— CATHEDRALS OF BANGOR, ST. ASAPH,
Llandaff, and St. David's. With Illustrations. Post 8vo. 15s.
———————— NORTHAMPTONSHIRE AND RUTLAND—
Northampton, Peterborough, Towcester, Daventry, Market Harborough, Kettering, Wallingborough, Thrapston, Stamford, Uppingham, Oakham. Maps. Post 8vo. 7s. 6d.
———————— DERBY, NOTTS, LEICESTER, STAFFORD,
Matlock, Bakewell, Chatsworth, The Peak, Buxton, Hardwick, Dove
Dale, Ashborne, Southwell, Mansfield, Retford, Burton, Belvoir, Melton
Mowbray, Wolverhampton, Lichfield, Walsall, Tamworth. Map.
Post 8vo. 9s.
———————— SHROPSHIRE AND CHESHIRE, Shrewsbury, Ludlow, Bridgnorth, Oswestry, Chester, Crewe, Alderley, Stockport,
Birkenhead. Maps and Plans. Post 8vo. 6s.
———————— LANCASHIRE, Warrington, Bury, Manchester,
Liverpool, Burnley, Clitheroe, Bolton, Blackburn, Wigan, Preston,
Rochdale, Lancaster, Southport, Blackpool, &c. Maps and Plans.
Post 8vo. 7s. 6d.
———————— YORKSHIRE, Doncaster, Hull, Selby, Beverley,
Scarborough, Whitby, Harrogate, Ripon, Leeds, Wakefield, Bradford,
Halifax, Huddersfield, Sheffield. Map and Plans. Post 8vo. 12s.
———————— CATHEDRALS of York, Ripon, Durham, Carlisle,
Chester, and Manchester. With 60 Illustrations. 2 Vols. Cr. 8vo. 21s.
———————— DURHAM AND NORTHUMBERLAND, Newcastle, Darlington, Gateshead, Bishop Auckland, Stockton, Hartlepool,
Sunderland, Shields, Berwick-on-Tweed, Morpeth, Tynemouth, Coldstream, Alnwick, &c. Map. Post 8vo. 9s.
———————— WESTMORLAND AND CUMBERLAND—Lancaster, Furness Abbey, Ambleside, Kendal, Windermere, Coniston,
Keswick, Grasmere, Ulswater, Carlisle, Cockermouth, Penrith, Appleby.
Map. Post 8vo.
※* MURRAY'S MAP OF THE LAKE DISTRICT, on canvas. 3s. 6d.
———————— SCOTLAND, Edinburgh, Melrose, Kelso, Glasgow,
Dumfries, Ayr, Stirling, Arran, The Clyde, Oban, Inverary, Loch
Lomond, Loch Katrine and Trossachs, Caledonian Canal, Inverness,
Perth, Dundee, Aberdeen, Braemar, Skye, Caithness, Ross, Sutherland, &c. Maps and Plans. Post 8vo. 9s.
———————— IRELAND, Dublin, Belfast, the Giant's Causeway, Donegal, Galway, Wexford, Cork, Limerick, Waterford, Killarney, Bantry, Glengariff, &c. Maps and Plans. Post 8vo. 10s.

HOME AND COLONIAL LIBRARY. A Series of Works adapted for all circles and classes of Readers, having been selected for their acknowledged Interest, and ability of the Authors. Post 8vo. Published at 2s. and 3s. 6d. each, and arranged under two distinctive heads as follows:—

CLASS A.
HISTORY, BIOGRAPHY, AND HISTORIC TALES.

1. SIEGE OF GIBRALTAR. By JOHN DRINKWATER. 2s.
2. THE AMBER-WITCH. By LADY DUFF GORDON. 2s.
3. CROMWELL AND BUNYAN. By ROBERT SOUTHEY. 2s.
4. LIFE OF SIR FRANCIS DRAKE. By JOHN BARROW. 2s.
5. CAMPAIGNS AT WASHINGTON. By REV. G. R. GLEIG. 2s.
6. THE FRENCH IN ALGIERS. By LADY DUFF GORDON. 2s.
7. THE FALL OF THE JESUITS. 2s.
8. LIVONIAN TALES. 2s.
9. LIFE OF CONDÉ. By LORD MAHON. 3s. 6d.
10. SALE'S BRIGADE. By REV. G. R. GLEIG. 2s.
11. THE SIEGES OF VIENNA. By LORD ELLESMERE. 2s.
12. THE WAYSIDE CROSS. By CAPT. MILMAN. 2s.
13. SKETCHES OF GERMAN LIFE. By SIR A. GORDON. 3s. 6d.
14. THE BATTLE OF WATERLOO. By REV. G. R. GLEIG. 3s. 6d.
15. AUTOBIOGRAPHY OF STEFFENS. 2s.
16. THE BRITISH POETS. By THOMAS CAMPBELL. 3s. 6d.
17. HISTORICAL ESSAYS. By LORD MAHON. 3s. 6d.
18. LIFE OF LORD CLIVE. By REV. G. R. GLEIG. 3s. 6d.
19. NORTH-WESTERN RAILWAY. By SIR F. B. HEAD. 2s
20. LIFE OF MUNRO. By REV. G. R. GLEIG. 3s. 6d.

CLASS B.
VOYAGES, TRAVELS, AND ADVENTURES.

1. BIBLE IN SPAIN. By GEORGE BORROW. 3s. 6d.
2. GYPSIES OF SPAIN. By GEORGE BORROW. 3s. 6d.
3 & 4. JOURNALS IN INDIA. By BISHOP HEBER. 2 Vols. 7s.
5. TRAVELS IN THE HOLY LAND. By IRBY and MANGLES. 2s.
6. MOROCCO AND THE MOORS. By J. DRUMMOND HAY. 2s.
7. LETTERS FROM THE BALTIC. By A LADY.
8. NEW SOUTH WALES. By MRS. MEREDITH. 2s.
9. THE WEST INDIES. By M. G. LEWIS. 2s.
10. SKETCHES OF PERSIA. By SIR JOHN MALCOLM. 3s. 6d.
11. MEMOIRS OF FATHER RIPA. 2s.
12 & 13. TYPEE AND OMOO. By HERMANN MELVILLE. 2 Vols. 7s.
14. MISSIONARY LIFE IN CANADA. By REV. J. ABBOTT. 2s.
15. LETTERS FROM MADRAS. By A LADY. 2s.
16. HIGHLAND SPORTS. By CHARLES ST. JOHN. 3s. 6d.
17. PAMPAS JOURNEYS. By SIR F. B. HEAD. 2s.
18. GATHERINGS FROM SPAIN. By RICHARD FORD. 3s. 6d.
19. THE RIVER AMAZON. By W. H. EDWARDS. 2s.
20. MANNERS & CUSTOMS OF INDIA. By REV. C. ACLAND. 2s.
21. ADVENTURES IN MEXICO. By G. F. RUXTON. 3s. 6d.
22. PORTUGAL AND GALICIA. By LORD CARNARVON. 3s. 6d.
23. BUSH LIFE IN AUSTRALIA. By REV. H. W. HAYGARTH. 2s.
24. THE LIBYAN DESERT. By BAYLE ST. JOHN. 2s.
25. SIERRA LEONE. By A LADY. 3s. 6d.

⁎ Each work may be had separately.

HOLLWAY (J. G.). A Month in Norway. Fcap. 8vo. 2s.

HONEY BEE. By Rev. Thomas James. Fcap. 8vo. 1s.

HOOK (Dean). Church Dictionary. 8vo. 16s.

—————— (Theodore) Life. By J. G. Lockhart. Fcap. 8vo. 1s.

HOPE (A. J. Beresford). Worship in the Church of England. 8vo. 9s., or, *Popular Selections from.* 8vo. 2s. 6d.

HORACE; a New Edition of the Text. Edited by Dean Milman. With 100 Woodcuts. Crown 8vo. 7s. 6d.

HOUGHTON'S (Lord) Monographs, Personal and Social. With Portraits. Crown 8vo. 10s. 6d.

—————— Poetical Works. *Collected Edition.* With Portrait. 2 Vols. Fcap. 8vo. 12s.

HOUSTOUN (Mrs.). Twenty Years in the Wild West of Ireland, or Life in Connaught. Post 8vo. 9s.

HUME (The Student's). A History of England, from the Invasion of Julius Cæsar to the Revolution of 1688. New Edition, revised, corrected, and continued to the Treaty of Berlin, 1878. By J. S. Brewer, M.A. With 7 Coloured Maps & 70 Woodcuts. Post 8vo. 7s. 6d.

HUTCHINSON (Gen.). Dog Breaking, with Odds and Ends for those who love the Dog and the Gun. With 40 Illustrations. Crown 8vo. 7s. 6d.

HUTTON (H. E.). Principia Græca; an Introduction to the Study of Greek. Comprehending Grammar, Delectus, and Exercise-book, with Vocabularies. *Sixth Edition.* 12mo. 3s. 6d.

HYMNOLOGY, Dictionary of. See Julian.

INDIA. [See Elphinstone, Hand-book, Temple.]

IRBY AND MANGLES' Travels in Egypt, Nubia, Syria, and the Holy Land. Post 8vo. 2s.

JAMESON (Mrs.). Lives of the Early Italian Painters— and the Progress of Painting in Italy—Cimabue to Bassano. With 50 Portraits. Post 8vo. 12s.

JAPAN. [See Bird, Mossman, Mounsey, Reed.]

JENNINGS (Louis J.). Field Paths and Green Lanes in Surrey and Sussex. Illustrations. Post 8vo. 10s. 6d.

—————— Rambles among the Hills in the Peak of Derbyshire and on the South Downs. With sketches of people by the way. With 28 Illustrations. Post 8vo. 12s.

JERVIS (Rev. W. H.). The Gallican Church, from the Concordat of Bologna, 1516, to the Revolution. With an Introduction. Portraits. 2 Vols. 8vo. 28s.

JESSE (Edward). Gleanings in Natural History. Fcp. 8vo. 3s. 6d.

JEX-BLAKE (Rev. T. W.). Life in Faith: Sermons Preached at Cheltenham and Rugb Fcap. 8vo. 3s. 6d.

JOHNSON'S (Dr. Samuel) Life. By James Boswell. Including the Tour to the Hebrides. Edited by Mr. Croker. 1 Vol. Royal 8vo. 12s.

JULIAN (Rev. John J.). A Dictionary of Hymnology. A Companion to Existing Hymn Books. Setting forth the Origin and History of the Hymns contained in the Principal Hymnals used by the Churches of England, Scotland, and Ireland, and various Dissenting Bodies, with Notices of their Authors. Post 8vo. [*In the Press.*

JUNIUS' Handwriting Professionally investigated. By Mr. Chabot, Expert. With Preface and Collateral Evidence, by the Hon. Edward Twisleton. With Facsimiles, Woodcuts, &c. 4to. £3 3s.

KERR (R. Malcolm). Student's Blackstone. A Systematic Abridgment of the entire Commentaries, adapted to the present state of the law. Post 8vo. 7s. 6d.

KING EDWARD VIth's Latin Grammar. 12mo. 3s. 6d.
—————— First Latin Book. 12mo. 2s. 6d.

KING (R. J.). Archæology, Travel and Art; being Sketches and Studies, Historical and Descriptive. 8vo. 12s.

KIRK (J. Foster). History of Charles the Bold, Duke of Burgundy. Portrait. 3 Vols. 8vo. 45s.

KIRKES' Handbook of Physiology. Edited by W. Morrant Baker, F.R.C.S. With 400 Illustrations. Post 8vo. 14s.

KUGLER'S Handbook of Painting.—The Italian Schools. Revised and Remodelled from the most recent Researches. By Lady Eastlake. With 140 Illustrations. 2 Vols. Crown 8vo. 30s.
—————— Handbook of Painting.—The German, Flemish, and Dutch Schools. Revised and in part re-written. By J. A. Crowe. With 60 Illustrations. 2 Vols. Crown 8vo. 24s.

LANE (E. W.). Account of the Manners and Customs of Modern Egyptians. With Illustrations. 2 Vols. Post 8vo. 12s.

LAWRENCE (Sir Geo.). Reminiscences of Forty-three Years' Service in India; including Captivities in Cabul among the Affghans and among the Sikhs, and a Narrative of the Mutiny in Rajputana. Crown 8vo. 10s. 6d.

LAYARD (A. H.). Nineveh and its Remains; a Popular Account of Researches and Discoveries amidst the Ruins of Assyria. With Illustrations. Post 8vo. 7s. 6d.
—————— Nineveh and Babylon; A Popular Account of Discoveries in the Ruins, with Travels in Armenia, Kurdistan and the Desert, during a Second Expedition to Assyria. With Illustrations. Post 8vo. 7s. 6d.

LEATHES (Stanley). Practical Hebrew Grammar. With the Hebrew Text of Genesis i.—vi., and Psalms i.—vi. Grammatical Analysis and Vocabulary. Post 8vo. 7s. 6d.

LENNEP (Rev. H. J. Van). Missionary Travels in Asia Minor. With Illustrations of Biblical History and Archæology. With Map and Woodcuts. 2 Vols. Post 8vo. 24s.
—————— Modern Customs and Manners of Bible Lands in Illustration of Scripture. With Coloured Maps and 300 Illustrations. 2 Vols. 8vo. 21s.

LESLIE (C. R.). Handbook for Young Painters. Illustrations. Post 8vo. 7s. 6d.
—————— Life and Works of Sir Joshua Reynolds. Portraits. 2 Vols. 8vo. 42s.

LETO (Pomponio). Eight Months at Rome during the Vatican Council. 8vo. 12s.

LETTERS From the Baltic. By a Lady. Post 8vo. 2s.
—————— Madras. By a Lady. Post 8vo. 2s.
—————— Sierra Leone. By a Lady. Post 8vo. 3s. 6d.

LEVI (Leone). History of British Commerce: and Economic Progress of the Nation, from 1763 to 1878. 8vo. 18s.

LEX SALICA; the Ten Texts with the Glosses and the Lex Emendata. Synoptically edited by J. H. Hessels. With Notes on the Frankish Words in the Lex Salica by H. Kern, of Leyden. 4to. 42s.

LIDDELL (Dean). Student's History of Rome, from the earliest Times to the establishment of the Empire. Woodcuts. Post 8vo. 7s. 6d.

LISPINGS from LOW LATITUDES; or, the Journal of the Hon. Impulsia Gushington. Edited by Lord Dufferin. With 24 Plates. 4to. 21s.

c 2

LIVINGSTONE (Dr.). First Expedition to Africa, 1840-56. Illustrations. Post 8vo. 7s. 6d.
—————— Second Expedition to Africa, 1858-64. Illustrations. Post 8vo. 7s. 6d.
—————— Last Journals in Central Africa, from 1865 to his Death. Continued by a Narrative of his last moments and sufferings. By Rev. Horace Waller. Maps and Illustrations. 2 Vols. 8vo. 15s.
—————— Personal Life. From his unpublished Journals and Correspondence. By Wm. G. Blaikie, D.D. With Map and Portrait. 8vo. 15s.
LIVINGSTONIA. Journal of Adventures in Exploring Lake Nyassa, and Establishing a Missionary Settlement there. By E. D. Young, R.N. Maps. Post 8vo. 7s. 6d.
LIVONIAN TALES. By the Author of "Letters from the Baltic." Post 8vo. 2s.
LOCKHART (J. G.). Ancient Spanish Ballads. Historical and Romantic. Translated, with Notes. Illustrations. Crown 8vo. 5s.
—————— Life of Theodore Hook. Fcap. 8vo. 1s.
LOUDON (Mrs.). Gardening for Ladies. With Directions and Calendar of Operations for Every Month. Woodcuts. Fcap. 8vo. 3s. 6d.
LYELL (Sir Charles). Principles of Geology; or, the Modern Changes of the Earth and its Inhabitants considered as illustrative of Geology. With Illustrations. 2 Vols. 8vo. 32s.
—————— Student's Elements of Geology. With Table of British Fossils and 600 Illustrations. Third Edition, Revised. Post 8vo. 9s.
—————— Life, Letters, and Journals. Edited by his sister-in-law, Mrs. Lyell. With Portraits. 2 Vols. 8vo. 30s.
—————— (K. M.). Geographical Handbook of Ferns. With Tables to show their Distribution. Post 8vo. 7s. 6d.
LYTTON (Lord). A Memoir of Julian Fane. With Portrait. Post 8vo. 5s.
McCLINTOCK (Sir L.). Narrative of the Discovery of the Fate of Sir John Franklin and his Companions in the Arctic Seas. With Illustrations. Post 8vo. 7s. 6d.
MACDOUGALL (Col.). Modern Warfare as Influenced by Modern Artillery. With Plans. Post 8vo. 12s.
MACGREGOR (J.). Rob Roy on the Jordan, Nile, Red Sea, Gennesareth, &c. A Canoe Cruise in Palestine and Egypt and the Waters of Damascus. With 70 Illustrations. Crown 8vo. 7s. 6d.
MAETZNER'S English Grammar. A Methodical, Analytical, and Historical Treatise on the Orthography, Prosody, Inflections, and Syntax. By Clair J. Grece, LL.D. 3 Vols. 8vo. 36s.
MAHON (Lord), see Stanhope.
MAINE (Sir H. Sumner). Ancient Law: its Connection with the Early History of Society, and its Relation to Modern Ideas. 8vo. 12s.
—————— Village Communities in the East and West. 8vo. 12s.
—————— Early History of Institutions. 8vo. 12s.
MALCOLM (Sir John). Sketches of Persia. Post 8vo. 3s. 6d.
MANSEL (Dean). Limits of Religious Thought Examined. Post 8vo. 8s. 6d.
—————— Letters, Lectures, and Reviews. 8vo. 12s.
MANUAL OF SCIENTIFIC ENQUIRY. For the Use of Travellers. Edited by Rev. R. Main. Post 8vo. 3s. 6d. (Published by order of the Lords of the Admiralty.)

MARCO POLO. The Book of Ser Marco Polo, the Venetian. Concerning the Kingdoms and Marvels of the East. A new English Version. Illustrated by the light of Oriental Writers and Modern Travels. By Col. Henry Yule. Maps and Illustrations. 2 Vols. Medium 8vo. 63s.

MARKHAM (Mrs.). History of England. From the First Invasion by the Romans. Woodcuts. 12mo. 3s. 6d.

———— History of France. From the Conquest by the Gauls. Woodcuts. 12mo. 3s. 6d.

———— History of Germany. From the Invasion by Marius. Woodcuts. 12mo. 3s. 6d.

———— (Clements R.). A Popular Account of Peruvian Bark and its introduction into British India. With Maps. Post 8vo. 14s.

MARRYAT (Joseph). History of Modern and Mediæval Pottery and Porcelain. With a Description of the Manufacture. Plates and Woodcuts. 8vo. 42s.

MARSH (G. P.). Student's Manual of the English Language. Edited with Additions. By Dr. Wm. Smith. Post 8vo. 7s. 6d.

MASTERS in English Theology. Lectures delivered at King's College, London, in 1877, by Eminent Divines. With Introduction by Canon Barry. Post 8vo. 7s. 6d.

MATTHIÆ'S Greek Grammar. Abridged by Blomfield, Revised by E. S. Crooke. 12mo. 4s.

MAUREL'S Character, Actions, and Writings of Wellington. Fcap. 8vo. 1s. 6d.

MAYO (Lord). Sport in Abyssinia; or, the Mareb and Tackazzee. With Illustrations. Crown 8vo. 12s.

MEADE (Hon. Herbert). Ride through the Disturbed Districts of New Zealand, with a Cruise among the South Sea Islands. With Illustrations. Medium 8vo. 12s.

MELVILLE (Hermann). Marquesas and South Sea Islands. 2 Vols. Post 8vo. 7s.

MEREDITH (Mrs. Charles). Notes and Sketches of New South Wales. Post 8vo. 2s.

MICHAEL ANGELO, Sculptor, Painter, and Architect. His Life and Works. By C. Heath Wilson. With Portrait. Illustrations and Index. 8vo. 15s.

MIDDLETON (Chas. H.) A Descriptive Catalogue of the Etched Work of Rembrandt, with Life and Introductions. With Explanatory Cuts. Medium 8vo. 31s. 6d.

MILLINGTON (Rev. T. S.). Signs and Wonders in the Land of Ham, or the Ten Plagues of Egypt, with Ancient and Modern Illustrations. Woodcuts. Post 8vo. 7s. 6d.

MILMAN'S (Dean) WORKS:—

History of the Jews, from the earliest Period down to Modern Times. 3 Vols. Post 8vo. 18s.

Early Christianity, from the Birth of Christ to the Abolition of Paganism in the Roman Empire. 3 Vols. Post 8vo. 18s.

Latin Christianity, including that of the Popes to the Pontificate of Nicholas V. 9 Vols. Post 8vo. 54s.

Handbook to St. Paul's Cathedral. Woodcuts. Crown 8vo. 10s. 6d.

Quinti Horatii Flacci Opera. Woodcuts. Sm. 8vo. 7s. 6d.

Fall of Jerusalem, Fcap. 8vo. 1s.

MILMAN'S (Capt. E. A.) Wayside Cross. Post 8vo. 2s.
———— [(Bishop, D.D.,) Life. With a Selection from his Correspondence and Journals. By his Sister. Map. 8vo. 12s.
MIVART (St. George). Lessons from Nature; as manifested in Mind and Matter. 8vo. 15s.
———— The Cat. An Introduction to the Study of Backboned Animals, especially Mammals. With 200 Illustrations. Medium 8vo. 30s.
MOORE (Thomas). Life and Letters of Lord Byron. *Cabinet Edition.* With Plates. 6 Vols. Fcap. 8vo. 18s.; *Popular Edition,* with Portraits. Royal 8vo. 7s. 6d.
MORESBY (Capt.), R.N. Discoveries in New Guinea, Polynesia, Torres Straits, &c., during the cruise of H.M.S. Basilisk. Map and Illustrations. 8vo. 15s.
MOSSMAN (Samuel). New Japan; the Land of the Rising Sun; its Annals during the past Twenty Years, recording the remarkable Progress of the Japanese in Western Civilisation. With Map. 8vo. 15s.
MOTLEY (J. L.). History of the United Netherlands: from the Death of William the Silent to the Twelve Years' Truce, 1609. Portraits. 4 Vols. Post 8vo. 6s. each.
———————— Life and Death of John of Barneveld, Advocate of Holland. With a View of the Primary Causes and Movements of the Thirty Years' War. Illustrations. 2 Vols. Post 8vo. 12s.
MOZLEY (Canon). Treatise on the Augustinian doctrine of Predestination. Crown 8vo. 9s.
MUIRHEAD (Jas.). The Vaux-de-Vire of Maistre Jean Le Houx, Advocate of Vire. Translated and Edited. With Portrait and Illustrations. 8vo. 21s.
MUNRO'S (General) Life and Letters. By Rev. G. R. Gleig. Post 8vo. 3s. 6d.
MURCHISON (Sir Roderick). Siluria; or, a History of the Oldest Rocks containing Organic Remains. Map and Plates. 8vo. 18s.
———————— Memoirs. With Notices of his Contemporaries, and Rise and Progress of Palæozoic Geology. By Archibald Geikie. Portraits. 2 Vols. 8vo. 30s.
MURRAY (A. S.). A History of Greek Sculpture, from the Earliest Times down to the Age of Pheidias. With Illustrations. Roy. 8vo. 21s.
MUSTERS' (Capt.) Patagonians; a Year's Wanderings over Untrodden Ground from the Straits of Magellan to the Rio Negro. Illustrations. Post 8vo. 7s. 6d.
NAPIER (Sir Wm.). English Battles and Sieges of the Peninsular War. Portrait. Post 8vo. 9s.
NAPOLEON at Fontainebleau and Elba. Journal of Occurrences and Notes of Conversations. By Sir Neil Campbell. Portrait. 8vo. 15s.
NARES (Sir George), R.N. Official Report to the Admiralty of the recent Arctic Expedition. Map. 8vo. 2s. 6d.
NAUTICAL ALMANAC (The). (*By Authority.*) 2s. 6d.
NAVY LIST. (Monthly and Quarterly.) Post 8vo.
NEW TESTAMENT. With Short Explanatory Commentary. By Archdeacon Churton, M.A., and the Bishop of St. David's. With 110 authentic Views &c. 2 Vols. Crown 8vo. 21s. *bound.*

NEWTH (SAMUEL). First Book of Natural Philosophy; an Introduction to the Study of Statics, Dynamics, Hydrostatics, Light, Heat, and Sound, with numerous Examples. Small 8vo. 8s. 6d.
—————— Elements of Mechanics, including Hydrostatics, with numerous Examples. Small 8vo. 8s. 6d.
—————— Mathematical Examples. A Graduated Series of Elementary Examples in Arithmetic, Algebra, Logarithms, Trigonometry, and Mechanics. Small 8vo. 8s. 6d.
NICOLAS (SIR HARRIS). Historic Peerage of England. Exhibiting the Origin, Descent, and Present State of every Title of Peerage which has existed in this Country since the Conquest. By WILLIAM COURTHOPE. 8vo. 30s.
NILE GLEANINGS. See STUART.
NIMROD, On the Chace—Turf—and Road. With Portrait and Plates. Crown 8vo. 5s. Or with Coloured Plates, 7s. 6d.
NORDHOFF (CHAS.). Communistic Societies of the United States; including Detailed Accounts of the Shakers, the Amana, Oneida, Bethell, Aurora, Icarian and other existing Societies. With 40 Illustrations. 8vo. 15s.
NORTHCOTE'S (SIR JOHN) Notebook in the Long Parliament. Containing Proceedings during its First Session, 1640. Edited, with a Memoir, by A. H. A. Hamilton. Crown 8vo. 9s.
OWEN (LIEUT.-COL.). Principles and Practice of Modern Artillery, including Artillery Material, Gunnery, and Organisation and Use of Artillery in Warfare. With Illustrations. 8vo. 15s.
OXENHAM (REV. W.). English Notes for Latin Elegiacs; designed for early Proficients in the Art of Latin Versification, with Prefatory Rules of Composition in Elegiac Metre. 12mo. 3s. 6d.
PAGET: (LORD GEORGE). The Light Cavalry Brigade in the Crimea. Containing Extracts from Journal and Correspondence. Map. Crown 8vo. 10s. 6d.
PALGRAVE (R. H. I.). Local Taxation of Great Britain and Ireland. 8vo. 6s.
PALLISER (MRS.). Mottoes for Monuments, or Epitaphs selected for General Use and Study. With Illustrations. Crown 8vo. 7s. 6d.
PARIS (DR.) Philosophy in Sport made Science in Earnest; or, the First Principles of Natural Philosophy inculcated by aid of the Toys and Sports of Youth. Woodcuts. Post 8vo. 7s. 6d.
PARKYNS' (MANSFIELD) Three Years' Residence in Abyssinia: with Travels in that Country. With Illustrations. Post 8vo. 7s. 6d.
PEEL'S (SIR ROBERT) Memoirs. 2 Vols. Post 8vo. 15s.
PENN (RICHARD). Maxims and Hints for an Angler and Chessplayer. Woodcuts. Fcap. 8vo. 1s.
PERCY (JOHN, M.D.). METALLURGY. Fuel, Wood, Peat, Coal, Charcoal, Coke. Fire-Clays. Illustrations. 8vo. 30s.
—————— Lead, including part of Silver. Illustrations. 8vo. 30s.
—————— Silver and Gold. Part I. Illustrations. 8vo. 30s.
PERRY (REV. CANON). Life of St. Hugh of Avalon, Bishop of Lincoln. Post 8vo. 10s. 6d.
—————— History of the English Church. See STUDENTS' Manuals.
PHILLIPS (SAMUEL). Literary Essays from "The Times." With Portrait. 2 Vols. Fcap. 8vo. 7s.
PIGAFETTA (FILIPPO). The Kingdom of Congo, and of the Surrounding Countries. Translated and edited by MARGARITE HUTCHINSON. With Preface by SIR T. F. BUXTON. Maps. 8vo. 10s. 6d.
POLLOCK (C. E.). A book of Family Prayers. Selected from the Liturgy of the Church of England. 16mo. 3s. 6d.

POPE'S (ALEXANDER) Works. With Introductions and Notes, by REV. WHITWELL ELWIN, and W. J. COURTHOPE. Vols. I., II., III., VI., VII., VIII. With Portraits. 8vo. 10s. 6d. each.

PORTER (REV. J. L.). Damascus, Palmyra, and Lebanon. With Travels among the Giant Cities of Bashan and the Hauran. Map and Woodcuts. Post 8vo. 7s. 6d.

PRAYER-BOOK (ILLUSTRATED), with Borders, Initials, Vignettes, &c. Edited, with Notes, by REV. THOS. JAMES. Medium 8vo. 18s. cloth ; 31s. 6d. calf ; 36s. morocco.

——— (THE CONVOCATION), with altered rubrics, showing the book if amended in conformity with the recommendations of the Convocations of Canterbury and York in 1879. Post 8vo. 6s.

PRINCESS CHARLOTTE OF WALES. A Brief Memoir. With Selections from her Correspondence and other unpublished Papers. By LADY ROSE WEIGALL. With Portrait. 8vo. 8s. 6d.

PRIVY COUNCIL JUDGMENTS in Ecclesiastical Cases relating to Doctrine and Discipline. With Historical Introduction, by G. C. BRODRICK and W. H. FREMANTLE. 8vo. 10s. 6d.

PSALMS OF DAVID. With Notes Explanatory and Critical by the Dean of Wells, Canon Elliott, and Canon Cook. Medium 8vo. 10s. 6d.

PUSS IN BOOTS. With 12 Illustrations. By OTTO SPECKTER. 16mo. 1s. 6d. Or coloured, 2s. 6d.

QUARTERLY REVIEW (THE). 8vo. 6s.

RAE (EDWARD). Country of the Moors. A Journey from Tripoli in Barbary to the Holy City of Kairwan. Map and Etchings. Crown 8vo. 12s.

——— The White Sea Peninsula. Journey to the White Sea, and the Kola Peninsula. With Map and 26 Illustrations. Crown 8vo. 15s.

RAMBLES in the Syrian Deserts. Post 8vo. 10s. 6d.

RASSAM (HORMUZD). British Mission to Abyssinia. Illustrations. 2 Vols. 8vo. 28s.

RAWLINSON'S (CANON) Herodotus. A New English Version. Edited with Notes and Essays. Maps and Woodcut. 4 Vols. 8vo. 48s.

——— Five Great Monarchies of Chaldæa, Assyria, Media, Babylonia, and Persia. With Maps and Illustrations. 3 Vols. 8vo. 42s.

——— (SIR HENRY) England and Russia in the East ; a Series of Papers on the Political and Geographical Condition of Central Asia. Map. 8vo. 12s.

REED (Sir E. J.) Iron-Clad Ships ; their Qualities, Performances, and Cost. With Chapters on Turret Ships, Iron-Clad Rams, &c. With Illustrations. 8vo. 12s.

——— Letters from Russia in 1875. 8vo. 5s.

——— Japan: Its History, Traditions, and Religions. With Narrative of a Visit in 1879. Illustrations. 2 Vols. 8vo. 28s.

REJECTED ADDRESSES (THE). By JAMES AND HORACE SMITH. Woodcuts. Post 8vo. 3s. 6d.; or Popular Edition, Fcap. 8vo. 1s.

REMBRANDT. See MIDDLETON.

REYNOLDS' (SIR JOSHUA) Life and Times. By C. R. LESLIE, R.A. and TOM TAYLOR. Portraits. 2 Vols. 8vo. 42s.

RICARDO'S (DAVID) Political Works. With a Notice of his Life and Writings. By J. R. M'CULLOCH. 8vo. 16s.

RIPA (FATHER). Thirteen Years at the Court of Peking. Post 8vo. 2s.

ROBERTSON (CANON). History of the Christian Church, from the Apostolic Age to the Reformation, 1517. 8 Vols. Post 8vo. 6s. each.

ROBINSON (REV. DR.). Biblical Researches in Palestine and the Adjacent Regions, 1838—52. Maps. 3 Vols. 8vo. 42s.

ROBINSON (WM.) Alpine Flowers for English Gardens. With 70 Illustrations. Crown 8vo. 7s. 6d.
—————— Sub-Tropical Garden. Illustrations. Small 8vo. 5s.
ROBSON (E. R.). SCHOOL ARCHITECTURE. Remarks on the Planning, Designing, Building, and Furnishing of School-houses Illustrations. Medium 8vo. 18s.
ROME (HISTORY OF). See GIBBON—LIDDELL—SMITH—STUDENTS'.
ROYAL SOCIETY CATALOGUE OF SCIENTIFIC PAPERS. 8 vols. 8vo. 20s. each. Half morocco, 28s. each.
RUXTON (GEO. F.). Travels in Mexico; with Adventures among Wild Tribes and Animals of the Prairies and Rocky Mountains. Post 8vo. 3s.6d
ST. HUGH OF AVALON, Bishop of Lincoln; his Life by G. G. PERRY, Canon of Lincoln. Post 8vo. 10s. 6d.
ST. JOHN (CHARLES). Wild Sports and Natural History of the Highlands of Scotland. Illustrated Edition. Crown 8vo. 15s. *Cheap Edition,* Post 8vo. 3s. 6d.
—————— (BAYLE) Adventures in the Libyan Desert. Post 8vo. 2s.
SALDANHA (DUKE OF). See CARNOTA.
SALE'S (SIR ROBERT) Brigade in Affghanistan. With an Account of the Defence of Jellalabad. By REV. G. R. GLEIG. Post 8vo. 2s.
SCEPTICISM IN GEOLOGY; and the Reasons for It. An assemblage of facts from Nature combining to refute the theory of "Causes now in Action." By VERIFIER. Woodcuts. Crown 8vo. 6s.
SCOTT (SIR GILBERT). Lectures on the Rise and Development of Mediæval Architecture. Delivered at the Royal Academy. With 400 Illustrations. 2 Vols. Medium 8vo. 42s.
SCHLIEMANN (DR. HENRY). Troy and Its Remains. A Narrative of Researches and Discoveries made on the Site of Ilium, and in the Trojan Plain. With 500 Illustrations. Medium 8vo. 42s.
—————— Ancient Mycenæ and Tiryns. With 500 Illustrations. Medium 8vo. 50s.
—————— Ilios; the City and Country of the Trojans, including all Recent Discoveries and Researches made on the Site of Troy and the Troad. With an Autobiography. With 2000 Illustrations. Imperial 8vo. 50s.
SCHOMBERG (GENERAL). The Odyssey of Homer, rendered into English blank verse, Books I—XII. 8vo. 12s.
SEEBOHM (HENRY). Siberia in Europe; a Naturalist's Visit to the Valley of the Petchora in N.E. Russia. With notices of Birds and their migrations. With Map and Illustrations. Crown 8vo. 14s.
SELBORNE (LORD). Notes on some Passages in the Liturgical History of the Reformed English Church. 8vo. 6s.
SHADOWS OF A SICK ROOM. Preface by Canon LIDDON. 16mo. 2s. 6d.
SHAH OF PERSIA'S Diary during his Tour through Europe in 1873. Translated from the Original. By J. W. REDHOUSE. With Portrait and Coloured Title. Crown 8vo. 12s.
SHAW (T. B.). Manual of English Literature. Post 8vo. 7s. 6d.
—————— Specimens of English Literature. Selected from the Chief Writers. Post 8vo. 7s. 6d.
SHAW (ROBERT). Visit to High Tartary, Yarkand, and Kashgar (formerly Chinese Tartary), and Return Journey over the Karakorum Pass. With Map and Illustrations. 8vo. 16s.
SIERRA LEONE; Described in Letters to Friends at Home. By A LADY. Post 8vo. 3s. 6d.
SIMMONS (CAPT.). Constitution and Practice of Courts-Martial. 8vo. 15s.

SMILES' (SAMUEL, LL.D.) WORKS :—
BRITISH ENGINEERS; from the Earliest Period to the death of the Stephensons. With Illustrations. 5 Vols. Crown 8vo. 7s. 6d. each.
LIFE OF GEORGE STEPHENSON. Post 8vo. 3s. 6d.
LIFE OF A SCOTCH NATURALIST (THOS. EDWARD). Illustrations. Crown 8vo. 10s. 6d.
LIFE OF A SCOTCH GEOLOGIST AND BOTANIST (ROBERT DICK). Illustrations. Crown 8vo. 12s.
HUGUENOTS IN ENGLAND AND IRELAND. Crown 8vo. 7s. 6d.
SELF-HELP. With Illustrations of Conduct and Perseverance. Post 8vo. 6s. Or in French, 5s.
CHARACTER. A Book of Noble Characteristics. Post 8vo. 6s.
THRIFT. A Book of Domestic Counsel. Post 8vo. 6s.
DUTY. With Illustrations of Courage, Patience, and Endurance. Post 8vo. 6s.
INDUSTRIAL BIOGRAPHY; or, Iron Workers and Tool Makers. Post 8vo. 6s.
BOY'S VOYAGE ROUND THE WORLD. Illustrations. Post 8vo. 6s.

SMITH (DR. GEORGE) Student's Manual of the Geography of India. Post 8vo.
——— Life of John Wilson, D.D. (Bombay), Missionary and Philanthropist. Portrait. Post 8vo. 9s.
——— (PHILIP). History of the Ancient World, from the Creation to the Fall of the Roman Empire, A.D. 476. 3 Vols. 8vo. 31s. 6d.

SMITH'S (DR. WM.) DICTIONARIES :—
DICTIONARY OF THE BIBLE; its Antiquities, Biography, Geography, and Natural History. Illustrations. 3 Vols. 8vo. 105s.
CONCISE BIBLE DICTIONARY. With 300 Illustrations. Medium 8vo. 21s.
SMALLER BIBLE DICTIONARY. With Illustrations. Post 8vo. 7s. 6d.
CHRISTIAN ANTIQUITIES. Comprising the History, Institutions, and Antiquities of the Christian Church. With Illustrations. 2 Vols. Medium 8vo. 3l. 13s. 6d.
CHRISTIAN BIOGRAPHY, LITERATURE, SECTS, AND DOCTRINES; from the Times of the Apostles to the Age of Charlemagne. Medium 8vo. Vols. I. & II. 31s. 6d. each. (To be completed in 4 Vols.)
GREEK AND ROMAN ANTIQUITIES. With 500 Illustrations. Medium 8vo. 28s.
GREEK AND ROMAN BIOGRAPHY AND MYTHOLOGY. With 600 Illustrations. 3 Vols. Medium 8vo. 4l. 4s.
GREEK AND ROMAN GEOGRAPHY. 2 Vols. With 500 Illustrations. Medium 8vo. 56s.
ATLAS OF ANCIENT GEOGRAPHY—BIBLICAL AND CLASSICAL. Folio. 6l. 6s.
CLASSICAL DICTIONARY OF MYTHOLOGY, BIOGRAPHY, AND GEOGRAPHY. 1 Vol. With 750 Woodcuts. 8vo. 18s.
SMALLER CLASSICAL DICTIONARY. With 200 Woodcuts. Crown 8vo. 7s. 6d.
SMALLER GREEK AND ROMAN ANTIQUITIES. With 200 Woodcuts. Crown 8vo. 7s. 6d.
COMPLETE LATIN-ENGLISH DICTIONARY. With Tables of the Roman Calendar, Measures, Weights, and Money. 8vo. 21s.
SMALLER LATIN-ENGLISH DICTIONARY. 12mo. 7s. 6d.

SMITH'S (Dr. Wm.) Dictionaries—*continued*.
    Copious and Critical English-Latin Dictionary. 8vo. 21*s*.
    Smaller English-Latin Dictionary. 12mo. 7*s*. 6*d*.
SMITH'S (Dr. Wm.) ENGLISH COURSE:—
    School Manual of English Grammar, with Copious Exercises. Post 8vo. 3*s*. 6*d*.
    Primary English Grammar. 16mo. 1*s*.
    Manual of English Composition. With Copious Illustrations and Practical Exercises. 12mo. 3*s*. 6*d*.
    Primary History of Britain. 12mo. 2*s*. 6*d*.
    School Manual of Modern Geography, Physical and Political. Post 8vo. 5*s*.
    A Smaller Manual of Modern Geography. 16mo. 2*s*. 6*d*.
SMITH'S (Dr. Wm.) FRENCH COURSE:—
    French Principia. Part I. A First Course, containing a Grammar, Delectus, Exercises, and Vocabularies. 12mo. 3*s*. 6*d*.
    Appendix to French Principia. Part I. Containing additional Exercises, with Examination Papers. 12mo. 2*s*. 6*d*.
    French Principia. Part II. A Reading Book, containing Fables, Stories, and Anecdotes, Natural History, and Scenes from the History of France. With Grammatical Questions, Notes and copious Etymological Dictionary. 12mo. 4*s*. 6*d*.
    French Principia. Part III. Prose Composition, containing a Systematic Course of Exercises on the Syntax, with the Principal Rules of Syntax. 12mo. [*In the Press.*
    Student's French Grammar. By C. Heron-Wall. With Introduction by M. Littré. Post 8vo. 7*s*. 6*d*.
    Smaller Grammar of the French Language. Abridged from the above. 12mo. 3*s*. 6*d*.
SMITH'S (Dr. Wm.) GERMAN COURSE:—
    German Principia. Part I. A First German Course, containing a Grammar, Delectus, Exercise Book, and Vocabularies. 12mo. 3*s*. 6*d*.
    German Principia. Part II. A Reading Book; containing Fables, Stories, and Anecdotes, Natural History, and Scenes from the History of Germany. With Grammatical Questions, Notes, and Dictionary. 12mo. 3*s*. 6*d*.
    Practical German Grammar. Post 8vo. 3*s*. 6*d*.
SMITH'S (Dr. Wm.) ITALIAN COURSE:—
    Italian Principia. An Italian Course, containing a Grammar, Delectus, Exercise Book, with Vocabularies, and Materials for Italian Conversation. By Signor Ricci, Professor of Italian at the City of London College. 12mo. 3*s*. 6*d*.
    Italian Principia. Part II. A First Italian Reading Book, containing Fables, Anecdotes, History, and Passages from the best Italian Authors, with Grammatical Questions, Notes, and a Copious Etymological Dictionary. By Signor Ricci. 12mo. 3*s*. 6*d*.
SMITH'S (Dr. Wm.) LATIN COURSE:—
    The Young Beginner's First Latin Book: Containing the Rudiments of Grammar, Easy Grammatical Questions and Exercises, with Vocabularies. Being a Stepping stone to Principia Latina, Part I. for Young Children. 12mo. 2*s*.
    The Young Beginner's Second Latin Book: Containing an easy Latin Reading Book, with an Analysis of the Sentences, Notes, and a Dictionary. Being a Stepping-stone to Principia Latina, Part II. for Young Children. 12mo. 2*s*.

SMITH'S (DR. WM.) LATIN COURSE—*continued*.

PRINCIPIA LATINA. Part I. First Latin Course, containing a Grammar, Delectus, and Exercise Book, with Vocabularies. 12mo. 3s. 6d.
⁎ In this Edition the Cases of the Nouns, Adjectives, and Pronouns are arranged both as in the ORDINARY GRAMMARS and as in the PUBLIC SCHOOL PRIMER, together with the corresponding Exercises.

APPENDIX TO PRINCIPIA LATINA. Part I.; being Additional Exercises, with Examination Papers. 12mo. 2s. 6d.

PRINCIPIA LATINA. Part II. A Reading-book of Mythology, Geography, Roman Antiquities, and History. With Notes and Dictionary. 12mo. 3s. 6d.

PRINCIPIA LATINA. Part III. A Poetry Book. Hexameters and Pentameters; Eclog. Ovidianæ; Latin Prosody. 12mo. 3s. 6d.

PRINCIPIA LATINA. Part IV. Prose Composition. Rules of Syntax with Examples, Explanations of Synonyms, and Exercises on the Syntax. 12mo. 3s. 6d.

PRINCIPIA LATINA. Part V. Short Tales and Anecdotes for Translation into Latin. 12mo. 3s.

LATIN-ENGLISH VOCABULARY AND FIRST LATIN-ENGLISH DICTIONARY FOR PHÆDRUS, CORNELIUS NEPOS, AND CÆSAR. 12mo. 3s. 6d.

STUDENT'S LATIN GRAMMAR. For the Higher Forms. Post 8vo. 6s.

SMALLER LATIN GRAMMAR. For the Middle and Lower Forms. 12mo. 3s. 6d.

TACITUS, Germania, Agricola, &c. With English Notes. 12mo. 3s. 6d.

SMITH'S (DR. WM.) GREEK COURSE:—

INITIA GRÆCA. Part I. A First Greek Course, containing a Grammar, Delectus, and Exercise-book. With Vocabularies. 12mo. 3s. 6d.

APPENDIX TO INITIA GRÆCA. Part I. Containing additional Exercises. With Examination Papers. Post 8vo. 2s. 6d.

INITIA GRÆCA. Part II. A Reading Book. Containing Short Tales, Anecdotes, Fables, Mythology, and Grecian History. 12mo. 3s. 6d.

INITIA GRÆCA. Part III. Prose Composition. Containing the Rules of Syntax, with copious Examples and Exercises. 12mo. 3s. 6d.

STUDENT'S GREEK GRAMMAR. For the Higher Forms. By CURTIUS. Post 8vo. 6s.

SMALLER GREEK GRAMMAR. For the Middle and Lower Forms. 12mo. 3s. 6d.

GREEK ACCIDENCE. 12mo. 2s. 6d.

PLATO, Apology of Socrates, &c. With Notes. 12mo. 3s. 6d.

SMITH'S (DR. WM.) SMALLER HISTORIES:—

SCRIPTURE HISTORY. Woodcuts. 16mo. 3s. 6d.
ANCIENT HISTORY. Woodcuts. 16mo. 3s. 6d.
ANCIENT GEOGRAPHY. Woodcuts. 16mo. 3s. 6d.
MODERN GEOGRAPHY. 16mo. 2s. 6d.
GREECE. Maps and Woodcuts. 16mo. 3s. 6d.
ROME. Maps and Woodcuts. 16mo. 3s. 6d.
CLASSICAL MYTHOLOGY. Woodcuts. 16mo. 3s. 6d.
ENGLAND. Maps and Woodcuts. 16mo. 3s. 6d.
ENGLISH LITERATURE. 16mo. 3s. 6d.
SPECIMENS OF ENGLISH LITERATURE. 16mo. 3s. 6d.

SOMERVILLE (Mary). Personal Recollections from Early Life to Old Age. Portrait. Crown 8vo. 12s.
———— Physical Geography. Portrait. Post 8vo. 9s.
———— Connexion of the Physical Sciences. Post 8vo. 9s.
———— Molecular & Microscopic Science. Illustrations. 2 Vols. Post 8vo. 21s.

SOUTH (John F.). Household Surgery; or, Hints for Emergencies. With Woodcuts. Fcap. 8vo. 3s. 6d.

SOUTHEY (Robt). Lives of Bunyan and Cromwell. Post 8vo. 2s.

STÄEL (Madame de). See Stevens.

STANHOPE'S (Earl) WORKS:—
  History of England from the Reign of Queen Anne to the Peace of Versailles, 1701-83. 9 vols. Post 8vo. 5s. each.
  Life of William Pitt. Portraits. 3 Vols. 8vo. 36s.
  Miscellanies. 2 Vols. Post 8vo. 13s.
  British India, from its Origin to 1783. Post 8vo. 3s. 6d.
  History of "Forty-Five." Post 8vo. 3s.
  Historical and Critical Essays. Post 8vo. 3s. 6d.
  French Retreat from Moscow, and other Essays. Post 8vo. 7s. 6d.
  Life of Belisarius. Post 8vo. 10s. 6d.
  Life of Condé. Post 8vo. 3s. 6d.
  Story of Joan of Arc. Fcap. 8vo. 1s.
  Addresses on Various Occasions. 16mo. 1s.

STANLEY'S (Dean) WORKS:—
  Sinai and Palestine. Maps. 8vo. 14s.
  Bible in the Holy Land; Extracted from the above Work. Woodcuts. Fcap. 8vo. 2s. 6d.
  Eastern Church. 8vo. 12s.
  Jewish Church. From the Earliest Times to the Christian Era. 3 Vols. 8vo. 38s.
  Church of Scotland. 8vo. 7s. 6d.
  Epistles of St. Paul to the Corinthians. 8vo. 18s.
  Life of Dr. Arnold. Portrait. 2 Vols. Cr. 8vo. 12s.
  Canterbury Cathedral. Illustrations. Post 8vo. 7s. 6d.
  Westminster Abbey. Illustrations. 8vo. 15s.
  Sermons during a Tour in the East. 8vo. 9s.
  ———— on Public Occasions, Preached in Westminster Abbey. 8vo.
  The Beatitudes, and Sermons Addressed to Children in Westminster Abbey. Fcap. 8vo.
  Memoir of Edward, Catherine, and Mary Stanley. Cr. 8vo. 9s.
  Christian Institutions. Essays on Ecclesiastical Subjects. 8vo. 12s.

STEPHENS (Rev. W. R. W.). Life and Times of St. John Chrysostom. A Sketch of the Church and the Empire in the Fourth Century. Portrait. 8vo. 12s.

STEVENS (Dr. A.). Madame de Stüel; a Study of her Life and Times. The First Revolution and the First Empire. Portraits. 2 Vols. Crown 8vo. 21s.

STRATFORD DE REDCLIFFE (LORD). The Eastern Question.
Being a Selection from his Writings during the last Five Years of his Life. With a Preface by Dean Stanley. With Map. 8vo. 9s.

STREET (G. E.). Gothic Architecture in Spain. Illustrations. Royal 8vo. 30s.

——————— Italy, chiefly in Brick and Marble. With Notes on North of Italy. Illustrations. Royal 8vo. 26s.

——— Rise of Styles in Architecture. With Illustrations. 8vo.

STUART (VILLIERS). Nile Gleanings: The Ethnology, History, and Art of Ancient Egypt, as Revealed by Paintings and Bas-Reliefs. With Descriptions of Nubia and its Great Rock Temples, 59 Coloured Illustrations, &c. Medium 8vo. 31s. 6d.

STUDENTS' MANUALS :—

OLD TESTAMENT HISTORY; from the Creation to the Return of the Jews from Captivity. Maps and Woodcuts. Post 8vo. 7s. 6d.

NEW TESTAMENT HISTORY. With an Introduction connecting the History of the Old and New Testaments. Maps and Woodcuts. Post 8vo. 7s. 6d.

EVIDENCES OF CHRISTIANITY. By REV. H. WACE. Post 8vo.

ECCLESIASTICAL HISTORY. The Christian Church. By PHILIP SMITH, B.A.
PART I.—First Ten Centuries. From its Foundation to the full establishment of the Holy Roman Empire and the Papal Power. Woodcuts. Post 8vo. 7s. 6d.
PART II.—The Middle Ages and the Reformation. Woodcuts. Post 8vo. 7s. 6d.

HISTORY OF THE ENGLISH CHURCH. By Canon PERRY.
First Period, from the planting of the Church in Britain to the Accession of Henry VIII. Post 8vo. 7s. 6d.

——————— Second Period, from the accession of Henry VIII. to the silencing of Convocation in the 18th Century. Post 8vo. 7s. 6d.

ANCIENT HISTORY OF THE EAST; Egypt, Assyria, Babylonia, Media, Persia, Asia Minor, and Phœnicia. Woodcuts. Post 8vo. 7s. 6d.

ANCIENT GEOGRAPHY. By Canon BEVAN. Woodcuts. Post 8vo. 7s. 6d.

HISTORY OF GREECE; from the Earliest Times to the Roman Conquest. By WM. SMITH, D.C.L. Woodcuts. Crown 8vo. 7s. 6d.
*** Questions on the above Work, 12mo. 2s.

HISTORY OF ROME; from the Earliest Times to the Establishment of the Empire. By DEAN LIDDELL. Woodcuts. Crown 8vo. 7s. 6d.

GIBBON'S DECLINE AND FALL OF THE ROMAN EMPIRE. Woodcuts. Post 8vo. 7s. 6d.

HALLAM'S HISTORY OF EUROPE during the Middle Ages. Post 8vo. 7s. 6d.

HISTORY OF MODERN EUROPE, from the end of the Middle Ages to the Treaty of Berlin, 1878. Post 8vo. [In the Press.

HALLAM'S HISTORY OF ENGLAND; from the Accession of Henry VII. to the Death of George II. Post 8vo. 7s. 6d.

HUME'S HISTORY OF ENGLAND from the Invasion of Julius Cæsar to the Revolution in 1688. Revised, corrected, and continued down to the Treaty of Berlin, 1878. By J. S. BREWER, M.A. With 7 Coloured Maps & 70 Woodcuts. Post 8vo. 7s. 6d.
*** Questions on the above Work, 12mo. 2s.

HISTORY OF FRANCE; from the Earliest Times to the Establishment of the Second Empire, 1852. By H. W. JERVIS. Woodcuts. Post 8vo. 7s. 6d.

ENGLISH LANGUAGE. By GEO. P. MARSH. Post 8vo. 7s. 6d.

ENGLISH LITERATURE. By T. B. SHAW, M.A. Post 8vo. 7s. 6d.

STUDENTS' MANUALS—continued.
SPECIMENS OF ENGLISH LITERATURE. By T. B. SHAW. Post 8vo. 7s. 6d.
MODERN GEOGRAPHY; Mathematical, Physical and Descriptive. By CANON BEVAN. Woodcuts. Post 8vo. 7s. 6d.
GEOGRAPHY OF INDIA. By Dr. GEORGE SMITH, LL.D. Post 8vo 7s. 6d. [*In the Press*
MORAL PHILOSOPHY. By WM. FLEMING. Post 8vo. 7s. 6d.
BLACKSTONE'S COMMENTARIES. By MALCOLM KERR. Post 8vo 7s. 6d.

SUMNER'S (BISHOP) Life and Episcopate during 40 Years. By Rev. G. H. SUMNER. Portrait. 8vo. 14s.

SWAINSON (CANON). Nicene and Apostles' Creeds; Their Literary History; together with some Account of "The Creed of St. Athanasius." 8vo. 16s.

SWIFT (JONATHAN). Life of. By HENRY CRAIK, B.A.

SYBEL (VON) History of Europe during the French Revolution, 1789—1795. 4 Vols. 8vo. 48s.

SYMONDS' (REV. W.) Records of the Rocks; or Notes on the Geology, Natural History, and Antiquities of North and South Wales, Siluria, Devon, and Cornwall. With Illustrations. Crown 8vo. 12s.

TALMUD. See BARCLAY; DEUTSCH.

TEMPLE (SIR RICHARD). India in 1880. With Maps. 8vo. 16s.
——— Men and Events of My Time in India. 8vo.

THIBAUT'S (ANTOINE) Purity in Musical Art. Translated from the German. With a prefatory Memoir by W. H. Gladstone, M.P. Post 8vo. 7s. 6d.

THIELMANN (BARON). Journey through the Caucasus to Tabreez, Kurdistan, down the Tigris and Euphrates to Nineveh and Babylon, and across the Desert to Palmyra. Translated by CHAS. HENEAGE. Illustrations. 2 Vols. Post 8vo. 18s.

THOMSON (ARCHBISHOP). Lincoln's Inn Sermons. 8vo. 10s. 6d.
——— Life in the Light of God's Word. Post 8vo. 5s.
——— Word, Work, & Will: Collected Essays. Crown 8vo. 9s.

TITIAN'S LIFE AND TIMES. With some account of his Family, chiefly from new and unpublished Records. By CROWE and CAVALCASELLE. With Portrait and Illustrations. 2 Vols. 8vo. 21s.

TOCQUEVILLE'S State of Society in France before the Revolution, 1789, and on the Causes which led to that Event. Translated by HENRY REEVE. 8vo. 14s.

TOMLINSON (CHAS.); The Sonnet; Its Origin, Structure, and Place in Poetry. With translations from Dante, Petrarch, &c. Post 8vo. 9s.

TOZER (REV. H. F.) Highlands of Turkey, with Visits to Mounts Ida, Athos, Olympus, and Pelion. 2 Vols. Crown 8vo. 24s.
——— Lectures on the Geography of Greece. Map. Post 8vo. 9s.

TRISTRAM (CANON). Great Sahara. Illustrations. Crown 8vo. 15s.
——— Land of Moab; Travels and Discoveries on the East Side of the Dead Sea and the Jordan. Illustrations. Crown 8vo. 15s.

TRURO (BISHOP OF). The Cathedral: its Necessary Place in the Life and Work of the Church. Crown 8vo. 6s.

TWENTY YEARS' RESIDENCE among the Greeks, Albanians, Turks, Armenians, and Bulgarians. By an ENGLISH LADY. 2 Vols. Crown 8vo. 21s.

TWINING (REV. THOS.). Records of a Life of Studious Retirement Being Selections from His Correspondence. 8vo. [*In the Press*

TWISS' (HORACE) Life of Lord Eldon. 2 Vols. Post 8vo. 21s.

TYLOR (E. B.) Researches into the Early History of Mankind, and Development of Civilization. 3rd Edition Revised. 8vo. 12s.

——————— Primitive Culture; the Development of Mythology, Philosophy, Religion, Art, and Custom. 2 Vols. 8vo. 24s.

VATICAN COUNCIL. See LETO.

VIRCHOW (PROFESSOR). The Freedom of Science in the Modern State. Fcap. 8vo. 2s.

WACE (REV. HENRY). The Gospel and its Witnesses: the Principal Facts in the Life of our Lord, and the Authority of the Evangelical Narratives. Crown 8vo.

WELLINGTON'S Despatches during his Campaigns in India, Denmark, Portugal, Spain, the Low Countries, and France. 8 Vols. 8vo. 20s. each.

——————— Supplementary Despatches, relating to India, Ireland, Denmark, Spanish America, Spain, Portugal, France, Congress of Vienna, Waterloo and Paris. 14 Vols. 8vo. 20s. each. *⁎* *An Index*. 8vo. 20s.

——————— Civil and Political Correspondence. Vols. I. to VIII. 8vo. 20s. each.

——————— Speeches in Parliament. 2 Vols. 8vo. 42s.

WHEELER (G.). Choice of a Dwelling. Post 8vo. 7s. 6d.

WHITE (W. H.). Manual of Naval Architecture, for the use of Naval Officers, Shipowners, Shipbuilders, and Yachtsmen. Illustrations. 8vo. 24s.

WHYMPER (EDWARD). The Ascent of the Matterhorn. With 100 Illustrations. Medium 8vo. 10s. 6d.

WILBERFORCE'S (BISHOP) Life of William Wilberforce. Portrait. Crown 8vo. 6s.

——————— (SAMUEL, LL.D.), Lord Bishop of Oxford and Winchester; his Life. By Canon ASHWELL, D.D., and R. G. WILBERFORCE. With Portraits and Woodcuts. Vols. I. and II. 8vo. 15s. each.

WILKINSON (SIR J. G.). Manners and Customs of the Ancient Egyptians, their Private Life, Laws, Arts, Religion, &c. A new edition. Edited by SAMUEL BIRCH, LL.D. Illustrations. 3 Vols. 8vo. 84s.

——————— Popular Account of the Ancient Egyptians. With 500 Woodcuts. 2 Vols. Post 8vo. 12s.

WILSON (JOHN, D.D.). See SMITH (GEO.).

WOOD'S (CAPTAIN) Source of the Oxus. With the Geography of the Valley of the Oxus. By COL. YULE. Map. 8vo. 12s.

WORDS OF HUMAN WISDOM. Collected and Arranged by E. S. With a Preface by CANON LIDDON. Fcap. 8vo. 3s. 6d.

WORDSWORTH'S (BISHOP) Greece; Pictorial, Descriptive, and Historical. With an Introduction on the Characteristics of Greek Art by GEORGE SCHARF, F.S.A. A Revised Edition, by H. F. TOZER, M.A., with Illustrations. Royal 8vo.

YORK (ARCHBISHOP OF). Collected Essays. Contents.—Synoptic Gospels. Death of Christ. God Exists. Worth of Life. Design in Nature. Sports and Pastimes. Emotions in Preaching. Defects in Missionary Work. Limits of Philosophical Enquiry. Crown 8vo. 9s.

YULE (COLONEL). Book of Marco Polo. Illustrated by the Light of Oriental Writers and Modern Travels. With Maps and 80 Plates. 2 Vols. Medium 8vo. 63s.

——————— A. F. A Little Light on Cretan Insurrection. Post 8vo. 2s. 6d.

www.ingramcontent.com/pod-product-compliance
Lightning Source LLC
Chambersburg PA
CBHW031937230426
43672CB00010B/1956